Biomimetics

Edited by Maki K. Habib
and César Martín-Gómez

Published in London, United Kingdom

IntechOpen

Supporting open minds since 2005

Biomimetics
http://dx.doi.org/10.5772/intechopen.91558
Edited by Maki K. Habib and César Martín-Gómez

Contributors
Manfred Hartbauer, Lei Nie, Amin Shavandi, Can Wang, Yaling Deng, Khurram Yousaf, Kunjie Chen, Muhammad Azam Khan, Khadeer Ahmed, Parveen Kaur Sidhu, Kiran Nehra, Kirti Rani, Mohamed Y. Zakaria, Ken Saito, Yuki Takei, Katsuyuki Morishita, Riku Tazawa, Maki K. Habib, Hashem Rizk

First published in London, United Kingdom, 2021 by IntechOpen
IntechOpen is the global imprint of INTECHOPEN LIMITED, registered in England and Wales, registration number: 11086078, 5 Princes Gate Court, London, SW7 2QJ, United Kingdom
Printed in Croatia

British Library Cataloguing-in-Publication Data
A catalogue record for this book is available from the British Library

Additional hard and PDF copies can be obtained from orders@intechopen.com

Biomimetics
Edited by Maki K. Habib and César Martín-Gómez
p. cm.
Print ISBN 978-1-83962-170-3
Online ISBN 978-1-83962-171-0
eBook (PDF) ISBN 978-1-83962-211-3

For EU product safety concerns:
IN TECH d.o.o., Prolaz Marije Krucifikse Kozulić 3, 51000 Rijeka, Croatia, info@intechopen.com or visit our website at intechopen.com.

Meet the editors

Maki K. Habib – Doctor of Engineering Science in Intelligent and Autonomous Robots, University of Tsukuba, Japan. He is a Full Professor of Robotics and Mechatronics at the Mechanical Engineering Department, School of Sciences and Engineering, The American University in Cairo, Egypt. His main areas of research interest are focused on: Autonomous Vehicles: Control and Navigation, Human adaptive and friendly Mechatronics, Service robots and Humanitarian demining, Autonomous Arial Vehicles and Autonomous Underwater Vehicles, Telecooperation, Distributed teleoperation, and Collaborative control, Flexible automation, Wireless sensor networks and Ambient intelligence, Biomimetic and biomedical robots, industry 4.0: IoTs, CPSs, and WSNs, Intelligent control.

César Martín Gómez (Ph.D. Architect) has been responsible for building services and energy systems in complex buildings in Spain since 2000. He has worked in I&S Ingenieros, in the Architecture Department in the Spanish Renewable Energies Center (CENER), and as Building Services and Energy Coordinator in Mangado & Asociados. Nowadays he works as a researcher and professor in the Department of Construction, Building Services and Structures at the University of Navarra (Spain).

Contents

Preface

The evolution of nature led to the resolve of many research and development challenges. Nature tested every field of science and engineering leading to inventions with creative solutions for everything that can work well, adapt and last long. Biomimetics that is associated with bio-inspiration represents an emerging area of research development that studies, assesses, and adopts from nature the best biological ideas and products in order to solve modern science and engineering problems. This includes models, elements, materials, processes, mechanisms, concepts systems, techniques, structures, seniors, etc. Biologically inspired approaches create a new reality with great potential.

This handbook includes 9 chapters that contribute to the state-of-art and up-to-date knowledge on research advancement of the biomimetic approach. The topics of chapters introduce research at the interface among biology, sciences, engineering, and technology. The editors through this book hope to build a line of transformative research directions based on the adaptation of creative design and smart solutions from natural models to technical models, using methodologies, algorithms, and solutions that are opening new dimensions and areas of development about how to solve efficiency requirements by understanding the evolution of solutions in nature. This creative-based thinking and processes enable researchers to think out of the boxes and open the way for new scientific challenges and achievements.

Maki K. Habib
Mechanical Engineering Department,
The American University in Cairo,
New Cairo, Egypt

César M. Gómez
Department of Construction,
Building Services and Structures at the University of Navarra,
Pamplona, Spain

Chapter 1

Bio-Inspired Hydrogels via 3D Bioprinting

Lei Nie, Can Wang, Yaling Deng and Amin Shavandi

Abstract

Many soft tissues of the human body such as cartilages, muscles, and ligaments are mainly composed of biological hydrogels possessing excellent mechanical properties and delicate structures. Nowadays, bio-inspired hydrogels have been intensively explored due to their promising potential applications in tissue engineering. However, the traditional manufacturing technology is challenging to produce the bio-inspired hydrogels, and the typical biological composite topologies of bio-inspired hydrogels are accessible completed using 3D bioprinting at micrometer resolution. In this chapter, the 3D bioprinting techniques used for the fabrication of bio-inspired hydrogels were summarized, and the materials used were outlined. This chapter also focuses on the applications of bio-inspired hydrogels fabricated using available 3D bioprinting technologies. The development of 3D bioprinting techniques in the future would bring us closer to the fabrication capabilities of living organisms, which would be widely used in biomedical applications.

Keywords: 3D bioprinting, hydrogels, biopolymers, tissue engineering, biomaterials

1. Introduction

The design of scaffold materials that can guide tissue regeneration is a very challenging goal [1]. In addition, to support and promote the growth and differentiation of specific cells, an ideal scaffold requires careful control of the material's structure in the range of nanometers to centimeters, and some natural materials with complex structure exist in nature, which provides ideas for the design of ideal scaffolds [2]. These natural materials, such as mammal bones, abalone pearl layers and fish scales, which are composed of multi-layer biominerals and biopolymers, have complex microstructure, which can control the crack growth and fracture in three-dimensional (3D) direction, producing much more strength and toughness than their constituent materials [3–5]. Jellyfish and sea anemones, with a water content of up to 90%, show that their gelatinous bodies exhibit exciting mechanical properties and are able to respond quickly to various environmental stimuli [6–8]. There are also some soft support tissues (such as tendons, ligaments, meniscus, and cartilage), showing softness, toughness and impact resistance [9]. Because of the beneficial properties of natural composite materials, the design of bionic materials has attracted significant attention. Bio-inspired material is considered as a kind of material inspired by nature or biology and then developed by simulating some characteristics [10], and usually, the bio-inspired materials provide better functions than synthetic materials [11].

However, there are still many limitations on the fabrication of bio-inspired materials using traditional material manufacturing technology because they cannot accurately control the distribution and spatial trend of micro-holes inside the materials, and it is challenging to produce the contour matching with natural materials [12, 13]. Recently, 3D bioprinting technology has become a promising tool for manufacturing materials with high-precision, which can overcome the limitations compare with the traditional methods, and finally can eventually produce complex and delicate biomimetic 3D structures. Also, 3D bioprinting technology realizes the automatic biological preparation of cell-laden structure through the layered deposition of bio-inks *in vitro* and *in vivo* [14]. In addition, 3D bioprinting technologies are controlled by computers and can be combined with medical imaging systems, such as computed tomography and magnetic resonance imaging (MRI), also combined with computer-aided design (CAD) and computer-aided manufacturing (CAM), to generate personalized structures organized in different length proportions [15, 16]. Compared with the classical tissue engineering methods, 3D bioprinting allows the direct manufacture of complex 3D structures containing spatial variations of biological materials, cells and biochemical substances with the same structure, which significantly improves the biological simulation level of the composition, structure and biochemical characteristics of cell niche in the human body [17]. The complexity of the resulting structure is not only related to the application of tissue regeneration but also to the development of cell biology, drug development, and disease research *in vitro* models [18].

In recent years, in tissue engineering development, many materials have been developed to meet the needs of 3D bioprinting. The most common 3D bioprinting materials are metals, engineering plastics, photosensitive resins, bioplastics and polymer hydrogels. The bio-inspired hydrogels are very similar to natural extracellular matrix (ECM) and display potential advantages in tissue engineering [19]. Bio-inspired hydrogel provides an adequate and porous microenvironment that allows good nutrition and oxygen to diffuse into the encapsulated cells and can be modified to guide cellular processes with various physical, chemical, and biological cues [20]. Besides, these hydrogels are usually non-toxic or low toxic and have good reproducibility. Next, the 3D bioprinting techniques used for the fabrication of bio-inspired hydrogels were summarized, and the materials used for 3D bioprinting were outlined. This chapter also focuses on the applications of bio-inspired hydrogels.

2. 3D bioprinting techniques

There are several available 3D bioprinting techniques for fabricating bio-inspired hydrogels, including inkjet bioprinting, laser-assisted bioprinting, extrusion bioprinting, and stereolithography, as shown in **Figure 1** [21].

2.1 Inkjet bioprinting

During the inkjet bioprinting process, biomaterials are selectively placed on the construction platform layer by layer until the required structure is formed. The first inkjet printers for bioprinting applications were improved versions of commercial two-dimensional ink printers [22]. For the inkjet bioprinting, the ink in the ink cartridge is replaced by biomaterials, and the paper is replaced by an electronically controlled lifting table to provide the control of the third dimension Z-axis in addition to the X-and Y-axes. The bioprinter based on inkjet printing technology is customized to process and print biomaterials with higher resolution, accuracy and speed [16]. Inkjet bioprinters use thermal or acoustic forces to spray droplets

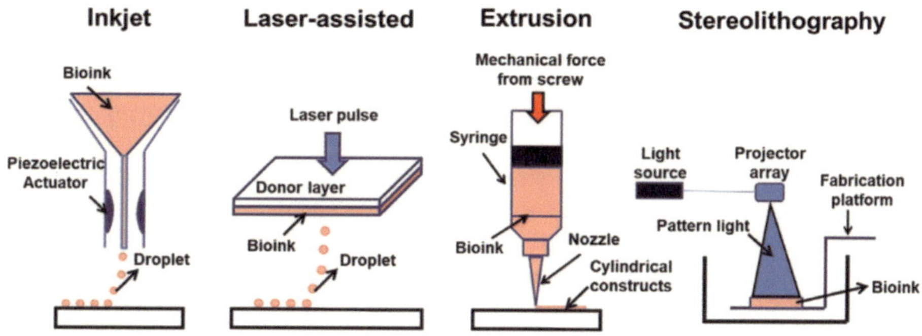

Figure 1.
Bioprinting techniques mainly include inkjet, laser-assisted, extrusion and stereolithography [21].

onto the substrate, which can support or form part of the final structure [23]. Thermal inkjet uses a heating element to induce the evaporation of a small volume of bioink in a reservoir, thereby forming and ejecting a small droplet. Therefore, in the printing process, this method keeps the cells at high temperature (300°C) for several microseconds (about 2 microseconds), which may lead to the formation of transient pores in the cell membrane [16]. Using the thermal inkjet printer, Solis et al., studied the effect of heat generated by the thermal ink-jet bio printer and found that the survival rate of Chinese hamster ovary (CHO) cells was 89% [24]. Such survival rate of cells could be greatly improved by using a piezoelectric inkjet printer, the generation and injection of droplets are realized by applying external voltage to control the mechanical deformation of piezoelectric transducer, which prevents the temperature from rising to the super physiological level [25]. Compaan et al. used alginate as the sacrificial material to prepare cell-supported silk fibroin hydrogels with a clear structure based on the piezoelectric inkjet 3D bioprinting system. The printed tubular structure has a diameter of 5 mm, a height of 2.5 or 5.0 mm and a thickness of about 400 microns. Moreover, the effect of citrate treatment on the printing was compared. The results showed that alginate removal and alginate removal could enable cells to extend and contact each other and form a cell network in the whole hydrogel [26].

The advantages of inkjet bioprinting mainly include: low cost due to its similar structure to commercial printers, high printing speed due to the ability of the print head to support parallel operation mode, and relatively high unit survival rate (usually from 80–90%) determined by many experimental results. However, the risks of cells and materials exposed to thermal and mechanical stresses, low droplet directionality, uneven droplet size, frequent nozzle plugging, and unreliable cell encapsulation have brought considerable limitations to the application in tissue engineering [27].

2.2 Laser-assisted bioprinting

The typical laser-assisted biological printing device include pulsed laser beams, focusing systems, and donor bands that respond to laser stimuli, consisting of glass covered with laser energy absorbing layers, and biomaterial layers (such as cells/hydrogel composite) prepared in liquid and receiving substrates for ribbons. The principle of laser-assisted bioprinting is to apply high-energy pulse laser (usually near-infrared laser) to the donor color band coated with bioink. This laser pulse evaporates a part of the donor layer, forms a high-pressure bubble on the interface of the bioink layer, and pushes the materials containing cells to the receiving substrate [16, 25, 28]. Compared with inkjet bioprinting, laser-assisted bioprinting can avoid the problem of jamming cell or material, also can avoid direct contact with the printer and biological ink at the same time. The non-contact biological printing method can choose much more types of ink, resulting in printing materials with wider range of viscosity [28].

The laser pulse energy, ECM thickness, and bioink viscosity can influence cell viability. The higher the laser energy is, the higher the cell death rate is, but the increase of membrane thickness and bioink viscosity will lead to an increase of cell viability. Guillotin et al. studied the effects of bioink viscosity, laser energy and printing speed on printing resolution. The microscale resolution and 5 kHz printing speed could be achieved, and the laser-assisted bioprinting could combine cells with ECM to produce soft tissue with high cell density *in vivo* [29]. Laser-assisted biological printing is considered as one of the most promising methods to fabricate engineered tissue because of its unique resolution, high throughput, high resolution, and high resolution, as well as the ability to produce heterogeneous tissue structures with high cell density [25]. However, compared with other bioprinting methods, the laser diode with high resolution and high intensity are expensive, and the control of the laser printing system is complex, which limit the application of this technology [28].

2.3 Extrusion bioprinting

The extrusion bioprinting can fabricate 3D cell carriers for tissue regeneration. The prepolymer solutions need to be prepared first, and almost all types of prepolymer solutions with different viscosities and aggregates with high cell density can be printed with extruded bioprinters [28]. Different from printing small droplets onto the platform, the extrusion bioprinting continuously deposit hydrogel filaments within a diameter of 150–300 microns to generate 3D structures. Common extrusion bioprinting method includes pneumatic, piston-driven, and screw-driven dispensing. In pneumatic dispensing, air pressure provides the required driving force, while in piston and screw-driven dispensing, vertical and rotating mechanical forces start printing respectively [30]. There are three main factors that decide the printability of extrusion bioprinting, mainly including the adjustability of viscosity, the bioink phase before extrusion, and the material-specific bio-manufacturing window [31]. Extrusion bioprinters have been used to produce various tissue types, such as aortic valves, branching vascular trees, in vitro drug movement and tumor models [32]. Although the manufacturing time may be prolonged for high-resolution complex structures, the structures have been manufactured from the clinically related tissue size to the microtissue in the microfluidic chamber. Furthermore, it is convenient to combine cells with bioactive agents, because that the heating process is not involved [33]. Compared with inkjet 3D bioprinting, extrusion bioprinters can achieve a continuous flow of biomaterials, thus achieving the simplicity of operation and a broader selection of biomaterials, including polymers, acellular matrices, cellular hydrogels, spheres and aggregates [34].

2.4 Stereolithography

Among all the bioprinting technologies, stereolithography (SLA) 3D bioprinting display much more advantages over extrusion or ink-jet bioprinting technology [28]. SLA is based on the polymerization of photosensitive polymers, and the digital mirror array controls the light band in the projection field to achieve selective crosslinking of each layer of the hydrogel prepolymer solution [35]. No matter how intricate a layer's pattern is, the printing time is the same because the whole pattern is projected on the printing plane. Therefore, the printer only needs a movable table in the vertical direction, which significantly simplifies the control of the printer. The cell encapsulated scaffold fabricated by the SLA system can achieve 100 μm resolution with printing time less than 1 hour, also maintain very high cell viability (90%) [36]. The above properties make SLA practical for fabricating delicate construct for tissue engineering. Arcaute et al. used composite lithography technology and two different molecular weight of polyethylene glycol (PEG) to prepare composite multilayer 3D structure of PEG hydrogel, and the properties of prepared hydrogel were influenced by photo-initiator and photosensitive polymer concentration. Besides, the prepared PEG hydrogel supports attachment, proliferation and differentiation of bovine chondrocytes, providing evidence for the applicability of resins for cartilage tissue engineering [37]. Valentin et al. prepared the sodium alginate precursor solution based on ion crosslinking, and different concentrations of cationic sources, such as barium carbonate, magnesium carbonate and calcium carbonate, and photo acid generator (PAG), diphenyliodonium nitrate were used, and the sodium alginate hydrogel was printed by SLA. The printed alginate hydrogel exhibited different mechanical and physical properties when crosslinked with two kinds of cations. The microstructures with variable height could be printed with optimized precursor formulations. Due to the high resolution, the 3D fabrication of natural and synthetic polyelectrolyte hydrogels via SLA enables lab-on-a-chip devices, soft sensors and actuators, and other biologically-inspired devices [38].

3. Polymers used for bio-inspired hydrogels

Hydrogels are considered as the gold standard materials for 3D bioprinting because they can provide a flexible and hydrated cross-linked network, similar to the natural extracellular matrix, in which cells can survive [39]. The polymers prepared for hydrogels can be classified into natural and synthetic polymers [40]. The natural polymers include alginate, chitosan, hyaluronic acid, gelatin, and so on, and the synthetic polymers mainly include polyacrylamide (PAAm), polyvinyl alcohol (PVA), polyethylene glycol (PEG), polylactic acid (PLA), and so on [41, 42].

3.1 Natural polymers

Most hydrogels prepared by natural polymers have the advantages of good hydrophilicity, good biocompatibility, specific enzymatic degradation, and contain various active functional groups and structural domains, and display better interaction with cells to promote cell proliferation and differentiation.

Alginate is extracted from alginate plants, is a kind of natural high molecular, composing of β-d-mannuronate (M) and α-l-guluronate (G). Alginate has been widely used in tissue engineering because of its advantages of abundant production, low price, good biocompatibility, and abundant functional groups, which are suitable for the preparation of bioink for 3D bioprinting [43, 44]. Alginate can react with $CaCO_3$ to release bivalent Ca^{2+} and then form an ionic crosslinking hydrogel bonded with -COO- on G unit of alginate G unit, to achieve the controllability of alginate ion crosslinking. The alginate hydrogel has high toughness and good mechanical properties, but the degradation rate of the alginate hydrogel is not controllable [45].

Chitosan is the product of deacetylation of chitin, which has a straight-chain structure and positive charge due to the presence of amino groups. Because of the useful biological function and biocompatibility, the degradation by microorganisms, chitosan has been widely concerned and applied in various industries [46]. The chitosan ink can be directly printed in air, and then the chitosan scaffold is refined by physical gelation. A chitosan hydrogel that satisfies both biocompatibility and mechanical properties has been obtained, and it has been confirmed that chitosan hydrogel can guide cell growth [47].

Gelatin is the hydrolysate of collagen, which contains many arginine-glycine-aspartic-acid (RGD) sequences and matrix metalloproteinase (MMP) target sequences, which enhance cell adhesion and cellular microenvironment remodeling respectively [48]. Because of biodegradability, biocompatibility, and low antigenicity, gelatin is attractive for bio-inspired hydrogel [49]. Lewis et al. used gelatin as a bioink to print into a specific 3D geometry using 3D bioprinting, which can regulate the biological processes of hepatocytes, enhance protein function, and facilitate cell proliferation and differentiation [50]. Another commonly used gelatin derivative is to acylate gelatin to form gelatin methacrylamide (GelMA) [51]. Zhou et al. used GelMA, N-(2-aminoethyl)-4-(4-(hydroxymethyl)-2-methoxy-5-nitrosophenoxy) butanamide linked hyaluronic acid (HA-NB) and photo-initiator lithium phenyl-2,4,6-trimethylbenzoylphosphinate (LAP) as biomimetic bioink to fabricated a bio-inspired 3D tissue construct via the digital light process (DLP)-based 3D bioprinting technology for ski regeneration (**Figure 2**) [52]. Bhise et al. used GelMA to carry out Hep G2/C3A cells to prepare biomimetic 3D liver structure hydrogel through bioprinting technology. A bionic human body chip of liver tissue was prepared by bioreactor. The toxicity response test of this chip in the test of acetaminophen is similar to that reported *in vivo* and other *in vitro* models, so this provides conceptual proof that the liver biomimetic human chip can be used in vitro drug toxicity screening experiments [53].

Hyaluronic acid (HA) is a kind of biocompatible non-sulfated glycosaminoglycan composed of N-acetylglucosamine and D-glucuronic acid repeated disaccharide units [54]. It is abundant in tissues including cartilage, neurons and skin. HA is of intrinsic biological importance because it binds to receptors such as CD44, can be degraded by oxidative species and hyaluronidase, and is related to the function and structure of development, wound healing and adult tissues. Because of biocompatibility, biodegradability, and natural biological function, HA hydrogels are widely used in various application fields [55]. Besides, the HA hydrogel can energize cell viability and promote osteoblasts to differentiate into cartilage. Unlike collagen and other proteins, the sequence of HA is different from species and its antigenicity is low, so it is especially promising as an injectable hydrogel.

Several other natural polymers, such as collagen, agarose, carrageenan, fibrin, heparin, chondroitin sulfate, cellulose, hemicellulose, lignin, and so on, could be used for hydrogels using 3D bioprinting [21]. However, natural hydrogels lack adequate mechanical properties, especially when implanted *in vivo* for a long time. Because of the uncontrollable swelling in physiological water environment, the mechanical stability of scaffolds tends to decrease. Thus, the chemical modification on natural polymers would be necessary to improve their printability as bioink, and the pending chemical groups after medication will improve the mechanical properties of construct after 3D bioprinting.

3.2 Synthetic polymers

The hydrogels fabricated using synthetic polymers have the advantages of long service life, strong water absorption, and high gel strength [41]. Polyacrylamide (PA) is a general designation of acrylamide homopolymer and copolymer. PA is a kind of water-soluble polymer, which has many amide groups in its structure and is easy to form hydrogen bond, so it has good stability and flocculation and is easy to be chemically modified. Ahn et al. grafted poly (N-isopropylacrylamide) (PNIPAAm) onto the framework of sodium alginate and synthesized sodium alginate PNIPAAm polymer micelles by self-assembly in aqueous solution, and the micelles could be used for the encapsulation of anticancer drug adriamycin [56]. Polyethylene glycol

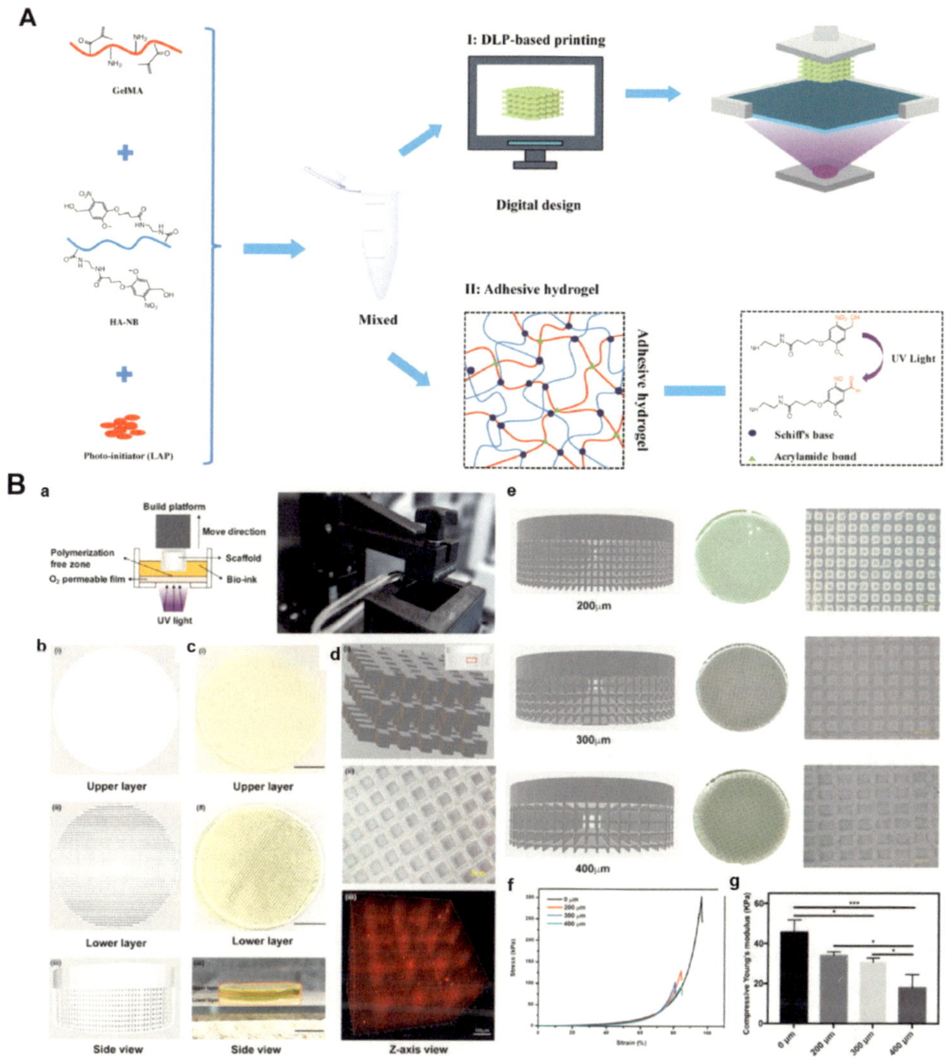

Figure 2.
(A) Fabrication of rapid gelation and tough GelMA/HA-NB/LAP hydrogel for DLP-based printing. (B) the skin analogous with sophisticated two-layer gel structure was fabricated via 3D bioprinting. (a) the bioink was printed with a layer-by-layer style using a DLP-based 3D printer. (b, c) the structure of native skin was displayed in CAD images. (d) the lower layer view of the scaffold was shown. (e) CAD images of different designed microchannel size and the printed products. (f) the elastic compressibility of products. (g) Compressive Young's modulus [52].

(PEG) is another synthetic polymer, and it has no toxicity and irritation, has good biocompatibility, and can be discharged from the body through the kidney. It has been widely used in the field of biomedicine [57]. Gao et al. constructed the polyethylene glycol diacrylate (PEGDA) hydrogel with uniform distribution of human mesenchymal stem cells (hMSCs) inside by simultaneous photopolymerization with commercial thermojet printers. hMSCs filled in 3D PEGDA hydrogel showed no deposition during culture and showed a chondrogenic phenotype [58]. Wang et al. prepared an injectable hydrogel through *in situ* Michael addition reaction between tetraniline-polyethylene glycol diacrylate (TA-PEG) and thiol hyaluronic acid (HA-SH), which was used to carry adipose-derived stem cells (ADSCs) [59].

Polylactic acid (PLA) is a kind of polymer, which is made of lactic acid as the primary raw material, and through polymerization, in which the performance can be adjusted by the structure [60]. Senatov et al. prepared PLA/hyaluronic acid

(HA) interconnected porous scaffold via a melt-wire method; the 3D printing technique avoided thermal degradation of PLA, the porosity and pore size of the scaffold could be well controlled. The porous PLA/HA scaffold with 15% HA has a considerable crack resistance and can work for a long time under the stress of 21 MPa, which was potential for bone tissue engineering applications [61].

Polyvinyl alcohol (PVA) is a synthetic water-soluble polymer, it has good biodegradation, biocompatibility, and no side effects on the human body [62]. PVA has been widely used in ophthalmology, wound dressing, artificial joint, and so on [42, 63]. Shi et al. prepared an injectable dynamic hydrogel using HA grafted with PVA and phenyl boric acid (PBA). The synthesized HA-PBA-PVA dynamic hydrogel has the reactive oxygen species reactivity and the scavenging activity of active oxygen. Furthermore, the hydrogel had good biocompatibility to the encapsulated neural precursor cells (NPC), and its ability to scavenge reactive oxygen species could protect the NPC cells from the damage of reactive oxygen species. The HA-PBA-PVA hydrogel could be used as bioink for 3D biological printing to prepare multilayer and cell loaded structures. The NPC cells showed good viability (85 ± 2% of living cells) after extrusion and maintained the excellent viability of 81 ± 2% of living cells after 3 days of culture. The results indicated that multifunctional injectable and ROS responsive self-healing HA-PBA-PVA dynamic hydrogels were expected to be candidates for 3D culture and 3D bioprinting [64].

Besides, there are also many other synthetic polymers for the fabrication of bio-inspired hydrogels, such as Pluronic and derivatives, PEG or polyethylene oxide (PEO) based block copolymers, poly(L-glutamic acid), poly(propylene fumarate), methoxy polyethylene glycol, and so on. Though, the synthetic polymers can precisely control their gel structure and properties and have better physical and chemical stability and more raw materials to prepare bio-inspired hydrogels. However, it is necessary to pay attention to the possible biocompatibility of unreacted monomers and residual initiators during the preparation of synthetic polymer materials, and the biocompatibility could be greatly improved via compositing or liking with natural polymers [65–67].

4. Applications of bio-inspired hydrogels using 3D bioprinting

4.1 Tissue regeneration

Tissue regeneration research is aim to develop substitute for damaged or diseased tissues or organs using principles of life science, engineering and medicine synergistically. It is crucial to fabricate the substitute as scaffolds, which is inspired by the natural 3D structure of tissue. The natural ECM regulates essential cellular functions, such as adhesion, migration, proliferation, differentiation and morphogenesis [68]. It is important of mimicking the ECM with dynamic nature using 3D bioprinting techniques, and the bio-inspired hydrogels via such techniques displayed potential applications in tissue regeneration, such as cartilage tissue, vascularized engineered tissue, bone tissue, skin regeneration, heart tissue, aortic valve conduits, muscle-tendon, and so on [69]. For example, Alexander et al. displayed a chemically and mechanically biomimetic filler-free bioink for 3D bioprinting of soft neural tissues, as shown in **Figure 3**. The thiolated Pluronic F-127, dopamine-conjugated (DC) gelatin, and DC hyaluronic acid were used as bioinks via a thiol-catechol reaction and photocuring; the storage modulus of the cured bioinks ranged from 6.7 to 11.7 kPa. The micro-extrusion 3D bioprinting was used to fabricate free-standing cell-laden tissue constructs. The Rodent Schwann

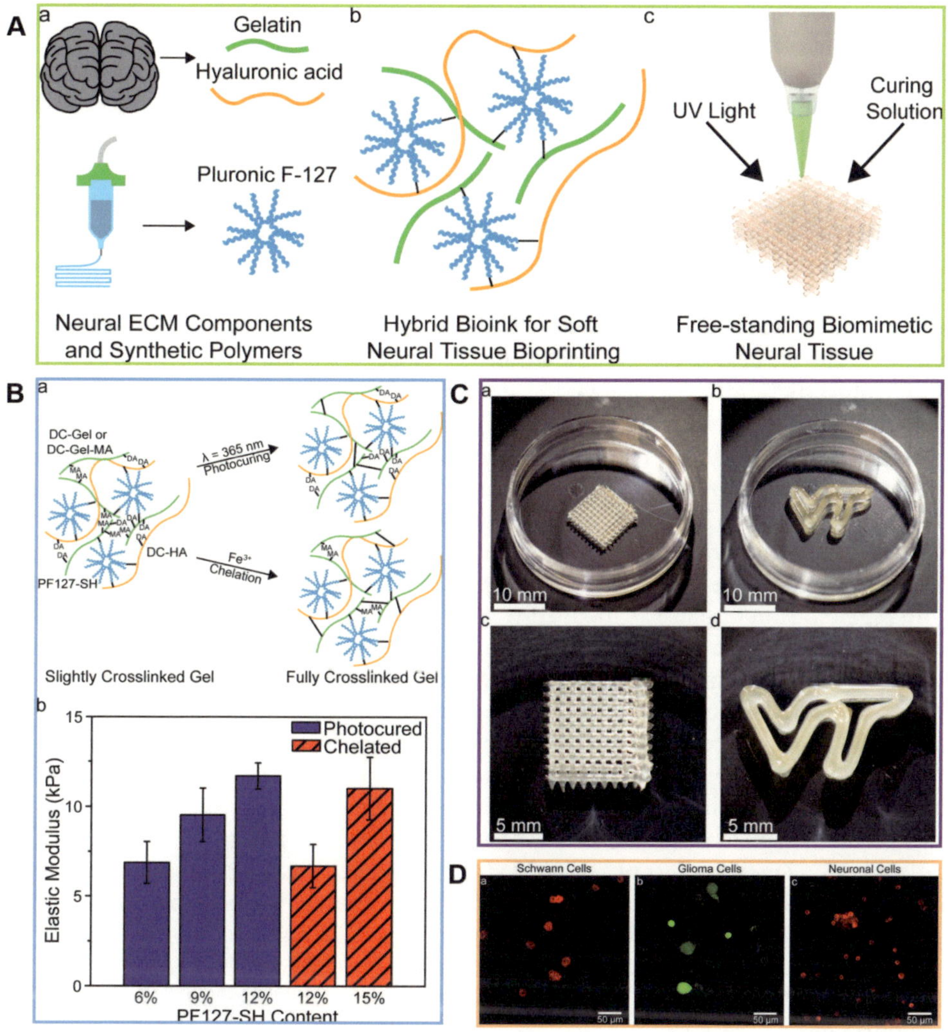

Figure 3.
(A) Native ECM components of neural tissue were combined with a synthetic polymer for microextrusion 3D bioprinting of soft, free-standing neural tissues. (B) Two curing pathways, including UV light exposure, and chelation of dopamine groups with iron (III), are shown to the formulation of photocuring containing methacrylated dopamine-conjugated gelatin. With the increase of PF127-SH content, the compressive properties of inks cured through UV exposure or chelation increased. **(C)** *Printed bioinks are shown. (D) Fluorescence micrographs of 3D bioprinted neural and glial tissue bioink containing rodent Schwann cells (a), human glioma cells (b), and rodent model neuronal cells (c) are shown at day 7 [70].*

cells, rodent neuronal cells, and human glioma cell-laden tissue constructs were printed and cultured over seven days and exhibited excellent viability, which has implications in micro physiological neural systems for neural tissue regenerative medicine [70]. Several works could be found in a recent study that focuses on the specific properties of bio-inspired hydrogels for tissue regeneration, such as high strength structures [30]. Also, the enhancement of printing resolution and versatility is vital for tissue regeneration. For example, the self-healing hydrogels were used to support the direct 3D bioprinting with high resolution by utilizing shear-thinning hydrogels, then the constructs could be printed in any direction [71]. The bio-inspired hydrogels could be accomplished via *in vitro* and *in vivo* 3D bioprinting as for tissue constructs, which are potential and convenient for clinic operation.

4.2 Wound dressing and wearable devices

The bio-inspired hydrogels via 3D bioprinting can be applied for wound dressing and wearable devices, which are considered as important applications, especially in recent years. Skin plays an essential role in protecting the body from external damages, such as abrasions, lacerations, and burns, and so on. The full-thickness defects of the dermis layers are the most challenging wounds to heal because of the limitation of self-repairing capability; thus, the skin regeneration of skin with skin appendages still remains a tough challenge [72]. 3D bioprinting is being applied to fabricate skin constructs using biomaterial scaffolds with or without cells, to address the need for skin tissues suitable for transplantation for wound healing therapy. The natural polymers, including cellulose, collagen and chitin, alginate, and hyaluronic acids are employed to synthesis skin constructs due to the favorable biocompatibility, biodegradation, low-toxicity or nontoxicity, high moisture content, high availability and mechanical stability [73]. Feifei et al. fabricated gelatin methacrylate (GelMA) based bioink to print functional living skin using DLP-based 3D printing (**Figure 2**), while the printed skin could promote skin regeneration and neovascularization via mimicking the physiological structure of natural skin [52].

Furthermore, the bio-inspired hydrogels could not only be functionalized on skin regeneration but also as medical wearable devices. The conductive hydrogels could be designed and fabricated to acquire electronic devices with conductive, capacitive, switching properties, image displaying, and motion sensing [74]. Meihong et al. developed conductive, healable, and self-adhesive hybrid network hydrogels based on conductive functionalized single-wall carbon nanotube (FSWCNT), PVA and polydopamine. The prepared hydrogel exhibits fast self-healing ability around 2 s, high self-healing efficiency of about 99%, and robust adhesiveness, which could be used for healable, adhesive, and soft human-motion sensors [75]. Zijian et al. synthesized a stretchable, self-healing and conductive hydrogel based on gelatin-enhanced hydrophobic association poly(acrylamide-*co*-dopamine) with lithium chloride via physical crosslinking including hydrogen bonding, hydrophobic association, and complexation effect. The hydrogels displayed the stretchability of 1150%, tensile strength of 112 kPa, flexibility and puncture resistance. Also, the hydrogels possess extraordinary conductive property and stable changes in resistance signals [76]. Furthermore, the organogel-hydrogel hybrids have been limelight due to that such kind of hybrids could mimic biological organisms with exceptional freezing tolerance, and thus could provide an advantageous skill to fabricate robust ionic skins [77]. Zhixing developed a series of lauryl acrylate-based polymeric organogels with high transparency, mechanical adaptability, adhesive capability, and self-healing properties; the prepared organogels were expected to provide insights to design the artificial human-like skins with unprecedented functionalities [78]. Due to the delicate structure can be accomplished using 3D bioprinting, bio-inspired hydrogel shows potential applications in medical wearable devices.

4.3 Pharmaceutical applications

The bio-inspired hydrogels could also be used in drug delivery system, such as protein carriers, anti-inflammatory drug carriers, in the pharmaceutical industry [79]. Rana et al. designed a magnetic natural hydrogel based on alginate, gelatin, and iron oxide magnetic nanoparticles as an efficient drug delivery system, the drug doxorubicin hydrochloride (DOX) was loaded, the anticancer activity against Hela cells could be regulated by the release of DOX from hydrogels [80]. Maling et al. provided a proof-of-concept of detoxification using a 3D-printed biomimetic nanocomposite construct in the hydrogel. A bio-inspired 3D detoxification device by installing polydiacetylene (PDA) nanoparticles in a 3D matrix was fabricated using dynamic optical projection stereolithography (DOPsL) technology; the nanoparticles could attract, capture and sense toxins, while the 3D matrix with a modified liver lobule microstructure allows toxins to be trapped efficiently [36]. The bio-inspired hydrogels via multi-materials 3D bioprinting can easy regulate the loading and release profiles of drugs, which show potentials as biomedicines.

5. Future outlook

The design paradigms shift from 2D to 3D has revolutionized the way of bio-inspired hydrogels for materials components, engineered constructs, *in vitro* disease modeling, medical wearable devices, and precision medicine. 3D bioprinting technology realizes to fabricate the delicate bio-inspired hydrogels with excellent properties and necessary signals to promote healing, tissue regeneration, therapeutics delivery, and health monitor in real-time. However, there are still some issues that need to be addressed in the near future (**Figure 4**). As the researchers begin to scale-up the production of bio-inspired hydrogels, new parameters during the fabrication need be met, such as the bioprinting speeds and resolutions, such parameters need to be simultaneously be increased to create constructs of clinic size. In the near future, it will be essential to develop microscale organ-on-a-chip, such as liver- and heart-on-a-chip, tumor-on-a-chip, etc., that integrate bio-inspired microenvironments with fluid flow inside hydrogels, also other dynamic physiological processes were well regulated by controlling the 3D bioprinting process. For example, the bio-inspired 3D culture in hydrogels could be employed to produce an *in vitro* model of Alzheimer's disease, providing a useful tool for the development of new therapeutics [82]. Future fabrication of bio-inspired hydrogels would be involved with multi-material 3D bioprinting, which provides the ability

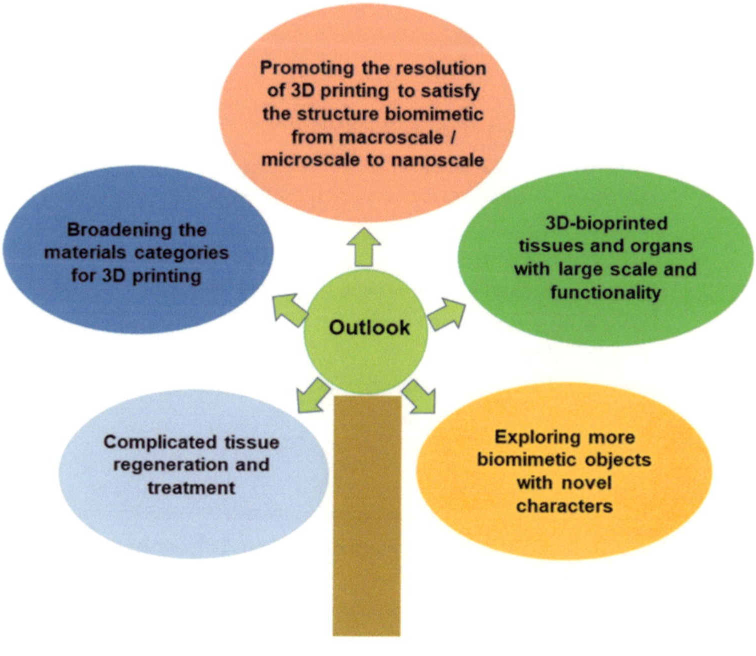

Figure 4.
The future outlook of 3D bioprinting for fabrication of bio-inspired tissues for tissue engineering applications [81].

to deliver growth factors, control cell adhesion, as well as the degradation rate in different regions of the printed constructs. In addition, 3D bioprinting technology needs to overcome vascularization challenge, which is considered a crucial factor in the synthesis of engineered constructs in tissue engineering.

6. Conclusions

The 3D bioprinting has changed the way bio-inspired hydrogels fabricated, and expanded the applications of bio-inspired hydrogels, including tissue regeneration, wound dressing, wearable devices, and pharmaceutical applications, and so on. In this chapter, the available 3D bioprinting techniques were described, the advantages and disadvantages of each printing technology were outlined. Then, the natural and synthetic polymers used for the fabrication of bio-inspired hydrogels via 3D bioprinting were introduced. The applications of bio-inspired hydrogels were focused. At last, the future outlook of bio-inspired hydrogels for tissue engineering were summarized. The bio-inspired hydrogels produced from 3D bioprinting still lacking sufficient clinical evidence, as more clinical trials evaluating bio-inspired hydrogels are still required.

Acknowledgements

This research was funded by the National Natural Science Foundation of China No. 31700840.

Conflict of interest

The authors declare no conflict of interest.

Abbreviations

ADSCs	adipose derived stem cells
CAD	computer-aided design
CAM	computer-aided manufacturing
CHO	Chinese hamster ovary
DLP	digital light process
DOPsL	dynamic optical projection stereolithography
DOX	doxorubicin hydrochloride
ECM	extracellular matrix
FSWCNT	functionalized single-wall carbon nanotube
GelMA	gelatin methacrylamide
HA	hyaluronic acid
HA-SH	thiol hyaluronic acid
hMSCs	human mesenchymal stem cells
LAP	lithium phenyl-2,4,6-trimethylbenzoylphosphinate
MMP	matrix metalloproteinase
MRI	magnetic resonance imaging
NB	N-(2-aminoethyl)-4-(4-(hydroxymethyl)-2-methoxy-5-nitrosophenoxy) butanamide
NPC	neural precursor cells

PA	polyacrylamide
PAAm	polyacrylamide
PAG	photo acid generator
PBA	phenyl boric acid
PDA	polydiacetylene
PEG	polyethylene glycol
PEGDA	polyethylene glycol diacrylate
PEO	polyethylene oxide
PLA	polylactic acid
PNIPAAm	poly (N-isopropylacrylamide)
PVA	polyvinyl alcohol
RGD	arginine-glycine-aspartic-acid
SLA	stereolithography
TA-PEGDA	tetraniline polyethylene glycol diacrylate

Author details

Lei Nie[1]*, Can Wang[1], Yaling Deng[2] and Amin Shavandi[3]*

1 College of Life Sciences, Xinyang Normal University, Xinyang, China

2 College of Intelligent Science and Control Engineering, Jinling Institute of Technology, Nanjing, China

3 BioMatter-Biomass Transformation Lab (BTL), Université Libre de Bruxelles, Avenue F.D. Roosevelt, Brussels, Belgium

*Address all correspondence to: nieleifu@yahoo.com, nielei@xynu.edu.cn and amin.shavandi@ulb.ac.be

References

[1] Guarino V, Gloria A, Raucci MG, De Santis R, Ambrosio L. Bio-inspired composite and cell instructive platforms for bone regeneration. International Materials Reviews. 2012;**57**(5):256-275

[2] Feilden E, Ferraro C, Zhang QH, Garcia-Tunon E, D'Elia E, Giuliani F, et al. 3D printing bioinspired ceramic composites. Scientific Reports. 2017;7

[3] Porter MM, Ravikumar N, Barthelat F, Martini R. 3D-printing and mechanics of bio-inspired articulated and multi-material structures. Journal of the Mechanical Behavior of Biomedical Materials. 2017;**73**:114-126

[4] Nie L, Li XC, Wang Z, Hu KH, Cai RH, Li P, et al. In vitro biomineralization on poly(vinyl alcohol)/biphasic calcium phosphate hydrogels. Bioinspired, Biomimetic and Nanobiomaterials. 2020;**9**(2):122-128

[5] Nie L, Deng YL, Li P, Hou RX, Shavandi A, Yang SF. Hydroxyethyl chitosan-reinforced polyvinyl alcohol/ biphasic calcium phosphate hydrogels for bone regeneration. ACS Omega. 2020;**5**(19):10948-10957

[6] Green JJ, Elisseeff JH. Mimicking biological functionality with polymers for biomedical applications. Nature. 2016;**540**(7633):386-394

[7] Wegst UGK, Bai H, Saiz E, Tomsia AP, Ritchie RO. Bioinspired structural materials. Nature Materials. 2015;**14**(1):23-36

[8] Ortiz C, Boyce MC. Materials science - bioinspired structural materials. Science. 2008;**319**(5866):1053-1054

[9] Hou RX, Wang YX, Han JL, Zhu YB, Zhang H, Zhang LB, et al. Structure and properties of PVA/silk fibroin hydrogels and their effects on growth behavior of various cell types. Materials Research Express. 2020;7(1).

[10] Munch E, Launey ME, Alsem DH, Saiz E, Tomsia AP, Ritchie RO. Tough, bio-inspired hybrid materials. Science. 2008;**322**(5907):1516-1520

[11] Wang W, Zhang YY, Liu WG. Bioinspired fabrication of high strength hydrogels from non-covalent interactions. Progress in Polymer Science. 2017;**71**:1-25

[12] Feng S, Liu JM, Ramalingam M. 3D printing of stem cell responsive ionically-crosslinked polyethylene glycol diacrylate/alginate composite hydrogels loaded with basic fibroblast growth factor for dental pulp tissue engineering: A preclinical evaluation in animal model. Journal of Biomaterials and Tissue Engineering. 2019;**9**(12):1635-1643

[13] Kim BS, Kwon YW, Kong JS, Park GT, Gao G, Han W, et al. 3D cell printing of in vitro stabilized skin model and in vivo pre-vascularized skin patch using tissue-specific extracellular matrix bioink: A step towards advanced skin tissue engineering. Biomaterials. 2018;**168**:38-53

[14] Xu T, Zhao WX, Zhu JM, Albanna MZ, Yoo JJ, Atala A. Complex heterogeneous tissue constructs containing multiple cell types prepared by inkjet printing technology. Biomaterials. 2013;**34**(1):130-139

[15] Vijayavenkataraman S, Yan WC, Lu WF, Wang CH, Fuh JYH. 3D bioprinting of tissues and organs for regenerative medicine. Advanced Drug Delivery Reviews. 2018;**132**:296-332

[16] Murphy SV, Atala A. 3D bioprinting of tissues and organs. Nature Biotechnology. 2014;**32**(8):773-785

[17] Dolatshahi-Pirouz A, Nikkhah M, Gaharwar AK, Hashmi B, Guermani E, Aliabadi H, et al. A combinatorial

cell-laden gel microarray for inducing osteogenic differentiation of human mesenchymal stem cells. Scientific Reports. 2014;**4**

[18] Horvath L, Umehara Y, Jud C, Blank F, Petri-Fink A, Rothen-Rutishauser B. Engineering an in vitro air-blood barrier by 3D bioprinting. Scientific Reports. 2015;**5**

[19] Liaw CY, Ji S, Guvendiren M. Engineering 3D Hydrogels for Personalized In Vitro Human Tissue Models. Advanced Healthcare Materials. 2018;7(4).

[20] Ng WL, Lee JM, Zhou MM, Yeong WY. Hydrogels for 3-D bioprinting-based tissue engineering. Woodh Publ Ser Biom. 2020:183-204

[21] Shavandi A, Hosseini S, Okoro OV, Nie L, Eghbali Babadi F, Melchels F. 3D Bioprinting of Lignocellulosic Biomaterials. Advanced Healthcare Materials.n/a(n/a):2001472.

[22] Xu T, Olson J, Zhao WX, Atala A, Zhu JM, Yoo JJ. Characterization of cell constructs generated with inkjet printing technology using in vivo magnetic resonance imaging. J Manuf Sci E-T Asme. 2008;130(2).

[23] Kim JI, Kim CS. Nanoscale resolution 3D printing with pin-modified electrified inkjets for tailorable Nano/macrohybrid constructs for tissue engineering. ACS Applied Materials & Interfaces. 2018;**10**(15):12390-12405

[24] Derby B. Bioprinting: Inkjet printing proteins and hybrid cell-containing materials and structures. Journal of Materials Chemistry. 2008;**18**(47):5717-5721

[25] Pereira RF, Bartolo PJ. 3D bioprinting of photocrosslinkable hydrogel constructs. Journal of Applied Polymer Science. 2015;132(48).

[26] Compaan AM, Christensen K, Huang Y. Inkjet bioprinting of 3D silk fibroin cellular constructs using sacrificial alginate. Acs Biomaterials Science & Engineering. 2017;**3**(8): 1519-1526

[27] Cui XF, Gao GF, Qiu YJ. Accelerated myotube formation using bioprinting technology for biosensor applications. Biotechnology Letters. 2013;**35**(3):315-321

[28] Mandrycky C, Wang ZJ, Kim K, Kim DH. 3D bioprinting for engineering complex tissues. Biotechnology Advances. 2016;**34**(4):422-434

[29] Guillotin B, Souquet A, Catros S, Duocastella M, Pippenger B, Bellance S, et al. Laser assisted bioprinting of engineered tissue with high cell density and microscale organization. Biomaterials. 2010;**31**(28): 7250-7256

[30] Derakhshanfar S, Mbeleck R, Xu KG, Zhang XY, Zhong W, Xing M. 3D bioprinting for biomedical devices and tissue engineering: A review of recent trends and advances. Bioactive Materials. 2018;**3**(2):144-156

[31] He Y, Yang FF, Zhao HM, Gao Q, Xia B, Fu JZ. Research on the printability of hydrogels in 3D bioprinting. Scientific Reports. 2016;**6**

[32] Duan B, Hockaday LA, Kang KH, Butcher JT. 3D bioprinting of heterogeneous aortic valve conduits with alginate/gelatin hydrogels. Journal of Biomedical Materials Research. Part A. 2013;**101**(5):1255-1264

[33] Jammalamadaka U, Tappa K. Recent Advances in Biomaterials for 3D Printing and Tissue Engineering. Journal of Functional Biomaterials. 2018;9(1).

[34] Cui HT, Nowicki M, Fisher JP, Zhang LG. 3D Bioprinting for Organ Regeneration. Advanced Healthcare Materials. 2017;6(1).

[35] Wang ZJ, Kumar H, Tian ZL, Jin X, Holzman JF, Menard F, et al. Visible light Photoinitiation of cell-adhesive gelatin Methacryloyl hydrogels for stereolithography 3D bioprinting. ACS Applied Materials & Interfaces. 2018;**10**(32):26859-26869

[36] Gou ML, Qu X, Zhu W, Xiang ML, Yang J, Zhang K, et al. Bio-inspired detoxification using 3D-printed hydrogel nanocomposites. Nature Communications. 2014;**5**

[37] Arcaute K, Mann BK, Wicker RB. Practical use of hydrogels in stereolithography for tissue engineering applications. Stereolithography: Materials, Processes and Applications. 2011:299-331

[38] Valentin TM, Leggett SE, Chen PY, Sodhi JK, Stephens LH, McClintock HD, et al. Stereolithographic printing of ionically-crosslinked alginate hydrogels for degradable biomaterials and microfluidics. Lab on a Chip. 2017;**17**(20):3474-3488

[39] Skardal A, Atala A. Biomaterials for integration with 3-D bioprinting. Annals of Biomedical Engineering. 2015;**43**(3):730-746

[40] Ahmed EM, Aggor FS, Awad AM, El-Aref AT. An innovative method for preparation of nanometal hydroxide superabsorbent hydrogel. Carbohydrate Polymers. 2013;**91**(2):693-698

[41] Ahmed EM. Hydrogel: Preparation, characterization, and applications: A review. Journal of Advanced Research. 2015;**6**(2):105-121

[42] Nie L, Zhang GH, Hou RX, Xu HP, Li YP, Fu J. Controllable promotion of chondrocyte adhesion and growth on PVA hydrogels by controlled release of TGF-beta 1 from porous PLGA microspheres. Colloids and Surfaces B: Biointerfaces. 2015;**125**:51-57

[43] Zhou Q J, Wang TW, Wang C, Wang Z, Yang YA, Li P, et al. Synthesis and characterization of silver nanoparticles-doped hydroxyapatite/alginate microparticles with promising cytocompatibility and antibacterial properties. Colloids and Surfaces A: Physicochemical and Engineering Aspects. 2020;**585**

[44] Rastogi P, Kandasubramanian B. Review of alginate-based hydrogel bioprinting for application in tissue engineering. Biofabrication. 2019;11(4).

[45] Nie L, Wang C, Hou RX, Li XY, Sun M, Suo JP, et al. Preparation and characterization of dithiol-modified graphene oxide nanosheets reinforced alginate nanocomposite as bone scaffold. Sn Applied Sciences. 2019;1(6).

[46] Nie L, Wu QY, Long HY, Hu KH, Li P, Wang C, et al. Development of chitosan/gelatin hydrogels incorporation of biphasic calcium phosphate nanoparticles for bone tissue engineering. Journal of Biomaterials Science-Polymer Edition. 2019;**30**(17):1636-1657

[47] Wu QH, Maire M, Lerouge S, Therriault D, Heuzey MC. 3D Printing of Microstructured and Stretchable Chitosan Hydrogel for Guided Cell Growth. Advanced Biosystems. 2017;1(6).

[48] Yue K, Trujillo-de Santiago G, Alvarez MM, Tamayol A, Annabi N, Khademhosseini A. Synthesis, properties, and biomedical applications of gelatin methacryloyl (GelMA) hydrogels. Biomaterials. 2015;**73**:254-271

[49] Klotz BJ, Gawlitta D, Rosenberg AJWP, Malda J, Melchels FPW. Gelatin-Methacryloyl hydrogels: Towards biofabrication-based tissue repair. Trends in Biotechnology. 2016;**34**(5):394-407

[50] Lewis PL, Green RM, Shah RN. 3D-printed gelatin scaffolds of differing pore geometry modulate hepatocyte

function and gene expression. Acta Biomaterialia. 2018;**69**:63-70

[51] Lin RZ, Chen YC, Moreno-Luna R, Khademhosseini A, Melero-Martin JM. Transdermal regulation of vascular network bioengineering using a photopolymerizable methacrylated gelatin hydrogel. Biomaterials. 2013; **34**(28):6785-6796

[52] Zhou FF, Hong Y, Liang RJ, Zhang XZ, Liao YG, Jiang DM, et al. Rapid printing of bio-inspired 3D tissue constructs for skin regeneration. Biomaterials. 2020;**258**

[53] Bhise NS, Manoharan V, Massa S, Tamayol A, Ghaderi M, Miscuglio M, et al. A liver-on-a-chip platform with bioprinted hepatic spheroids. Biofabrication. 2016;8(1).

[54] Larraneta E, Henry M, Irwin NJ, Trotter J, Perminova AA, Donnelly RF. Synthesis and characterization of hyaluronic acid hydrogels crosslinked using a solvent-free process for potential biomedical applications. Carbohydrate Polymers. 2018;**181**:1194-1205

[55] Highley CB, Prestwich GD, Burdick JA. Recent advances in hyaluronic acid hydrogels for biomedical applications. Current Opinion in Biotechnology. 2016;**40**:35-40

[56] Ahn DG, Lee J, Park SY, Kwark YJ, Lee KY. Doxorubicin-loaded alginate-g-poly(N-isopropylacrylamide) micelles for cancer imaging and therapy. ACS Applied Materials & Interfaces. 2014;**6**(24):22069-22077

[57] Zou P, Suo JP, Nie L, Feng SB. Temperature-responsive biodegradable star-shaped block copolymers for vaginal gels. Journal of Materials Chemistry. 2012;**22**(13):6316-6326

[58] Gao GF, Hubbell K, Schilling AF, Dai GH, Cui XF. Bioprinting Cartilage Tissue from Mesenchymal Stem Cells and PEG Hydrogel. 3d Cell Culture: Methods and Protocols. 2017;1612:391-8.

[59] Wang W, Tan BY, Chen JR, Bao R, Zhang XR, Liang S, et al. An injectable conductive hydrogel encapsulating plasmid DNA-eNOs and ADSCs for treating myocardial infarction. Biomaterials. 2018;**160**:69-81

[60] Inkinen S, Hakkarainen M, Albertsson AC, Sodergard A. From lactic acid to poly(lactic acid) (PLA): Characterization and analysis of PLA and its precursors. Biomacromolecules. 2011;**12**(3):523-532

[61] Senatov FS, Niaza KV, Stepashkin AA, Kaloshkin SD. Low-cycle fatigue behavior of 3d-printed PLA-based porous scaffolds. Compos Part B Eng. 2016;**97**:193-200

[62] Wang TW, Zhang F, Zhao R, Wang C, Hu KH, Sun Y, et al. Polyvinyl alcohol/sodium alginate hydrogels incorporated with silver nanoclusters via Green tea extract for antibacterial applications. Designed Monomers and Polymers. 2020;**23**(1):118-133

[63] Nie L, Chen D, Suo JP, Zou P, Feng SB, Yang Q, et al. Physicochemical characterization and biocompatibility in vitro of biphasic calcium phosphate/polyvinyl alcohol scaffolds prepared by freeze-drying method for bone tissue engineering applications. Colloids and Surfaces B: Biointerfaces. 2012;**100**:169-176

[64] Shi W, Hass B, Kuss MA, Zhang HP, Ryu SJ, Zhang DZ, et al. Fabrication of versatile dynamic hyaluronic acid-based hydrogels. Carbohydrate Polymers. 2020;**233**

[65] Li HJ, Tan C, Li L. Review of 3D printable hydrogels and constructs. Materials & Design. 2018;**159**:20-38

[66] Zou P, Suo JP, Nie L, Feng SB. Temperature-sensitive biodegradable

mixed star-shaped block copolymers hydrogels for an injection application. Polymer. 2012;**53**(6):1245-1257

[67] Nie L, Zou P, Feng SB, Suo JP. Temperature-sensitive star-shaped block copolymers hydrogels for an injection application: Phase transition behavior and biocompatibility. Journal of Materials Science-Materials in Medicine. 2013;**24**(3):689-700

[68] Dvir T, Timko BP, Kohane DS, Langer R. Nanotechnological strategies for engineering complex tissues. Nature Nanotechnology. 2011;**6**(1):13-22

[69] Antich C, de Vicente J, Jimenez G, Chocarro C, Carrillo E, Montanez E, et al. Bio-inspired hydrogel composed of hyaluronic acid and alginate as a potential bioink for 3D bioprinting of articular cartilage engineering constructs. Acta Biomaterialia. 2020;**106**:114-123

[70] Haring AP, Thompson EG, Tong YX, Laheri S, Cesewski E, Sontheimer H, et al. Process- and bio-inspired hydrogels for 3D bioprinting of soft free-standing neural and glial tissues. Biofabrication. 2019;11(2).

[71] Highley CB, Rodell CB, Burdick JA. Direct 3D Printing of Shear-Thinning Hydrogels into Self-Healing Hydrogels. Advanced Materials. 2015;27(34):5075–+.

[72] Shi LY, Zhao YN, Xie QF, Fan CX, Hilborn J, Dai JW, et al. Moldable Hyaluronan Hydrogel Enabled by Dynamic Metal-Bisphosphonate Coordination Chemistry for Wound Healing. Advanced Healthcare Materials. 2018;7(5).

[73] He P, Zhao JN, Zhang JM, Li B, Gou ZY, Gou ML, et al. Bioprinting of skin constructs for wound healing. Burns & Trauma. 2018;**6**

[74] Tai YL, Mulle M, Ventura IA, Lubineau G. A highly sensitive, low-cost, wearable pressure sensor based on conductive hydrogel spheres. Nanoscale. 2015;**7**(35):14766-14773

[75] Liao MH, Wan PB, Wen JR, Gong M, Wu XX, Wang YG, et al. Wearable, Healable, and Adhesive Epidermal Sensors Assembled from Mussel-Inspired Conductive Hybrid Hydrogel Framework. Advanced Functional Materials. 2017;27(48).

[76] Gao ZJ, Li YF, Shang XL, Hu W, Gao GH, Duan LJ. Bio-inspired adhesive and self-healing hydrogels as flexible strain sensors for monitoring human activities. Materials Science and Engineering: C. 2020;**106**

[77] Gao HN, Zhao ZG, Cai YD, Zhou JJ, Hua WD, Chen L, et al. Adaptive and freeze-tolerant heteronetwork organohydrogels with enhanced mechanical stability over a wide temperature range. Nature Communications. 2017;**8**

[78] Zhang ZX, Wang L, Yu HT, Zhang F, Tang L, Feng YY, et al. Highly transparent, self-healable, and adhesive Organogels for bio-inspired intelligent ionic skins. ACS Applied Materials & Interfaces. 2020;**12**(13):15657-15666

[79] Kim S, Park CB. Bio-inspired synthesis of minerals for energy, environment, and medicinal applications. Advanced Functional Materials. 2013;**23**(1):10-25

[80] Jahanban-Esfahlan R, Derakhshankhah H, Haghshenas B, Massoumi B, Abbasian M, Jaymand M. A bio-inspired magnetic natural hydrogel containing gelatin and alginate as a drug delivery system for cancer chemotherapy. International Journal of Biological Macromolecules. 2020;**156**:438-445

[81] Li T, Chang J, Zhu YF, Wu CT. 3D printing of bioinspired biomaterials

for tissue regeneration. Advanced Healthcare Materials. 2020

[82] Choi SH, Kim YH, Hebisch M, Sliwinski C, Lee S, D'Avanzo C, et al. A three-dimensional human neural cell culture model of Alzheimer's disease. Nature. 2014;**515**(7526):274-U93

Chapter 2

Active Gaits Generation of Quadruped Robot Using Pulse-Type Hardware Neuron Models

Yuki Takei, Katsuyuki Morishita, Riku Tazawa and Ken Saito

Abstract

In this chapter, the authors will propose the active gait generation of a quadruped robot. We developed the quadruped robot system using self-inhibited pulse-type hardware neuron models (P-HNMs) as a solution to elucidate the gait generation method. We feedbacked pressures at the robot system's each foot to P-HNM and varied the joints' angular velocity individually. We experimented with making the robot walk from an upright position on a flat floor. As a result of the experiment, we confirmed that the robot system spontaneously generates walk gait and trot gait according to the moving speed. Also, we clarified the process by which the robot actively generates gaits from the upright state. These results suggest that animals may generate gait using a similarly simple method because P-HNM mimics biological neurons' function. Furthermore, it shows that our robot system can generate gaits adaptively and quite easily.

Keywords: gait generation, interlimb coordination, pulse-type hardware neuron model (P-HNM), locomotion, quadruped robot, gait pattern

1. Introduction

Improvements in computer processing capability have realized advanced mobile robots [1, 2]. However, there is still no legged robot that can move flexibly enough to change our lives in a big way. One of the reasons for this is challenging to act autonomously for current control methods to instantly adapt to various events occurring in the robot's surroundings. The realization of the autonomous robot needs a higher sensory information processing system. Also, increasing the number of sensors or shorten the interval acquisition of the sensor information requires high-speed information processing. On the other hand, animals can easily act autonomously. The significant difference between robots and animals in deciding how they should act is to process all information by their brains or not. Mimicking the animal's biological system can be useful for realizing simple robot control [3, 4]. For example, the typical actions that animals unconsciously generate are respiration, chewing, and walking [5, 6]. The elucidation of generating walking action may solve the problem of current control methods for legged robots. The ordinary quadrupeds as legged animals have several locomotion patterns (gaits) [7–9]. Neurophysiology experiments have provided many insights into the characteristics and kinematics in gait [10–18]. The finding that horses move efficiently by switching their gait to suit the situation is essential for the engineering application of their gait generation mechanisms [19]. Researchers also examined how quadrupeds generate gait [19–22]. The theory that quadruped animals unconsciously generate gaits by the interaction between the central pattern generator (CPG) and sensory inputs is widely accepted [23–26]. The variety of animals' functions makes it difficult to use their bodies to identify the essential elements required to generate gait. Although there is much discussion on the animal's gait generating mechanisms, most of it is still unclear [27, 28].

Researchers have attempted to realize CPG in engineering and use modeled CPGs to control robots [29–35]. These studies have succeeded in using the CPG models to generate locomotion, which was previously calculated by the processor [32–35]. However, how does an animal's CPGs create the gait according to the robot's surroundings is unclear. It is necessary to examine a method for generating gait employing body structure to identify the essential elements required for entities to generate gait.

Research using a biped machine with passive joints revealed that the biped machine generates a gait pattern without a control mechanism when placed on a shallow slope [36]. Another research using a quadruped machine revealed that it generates quadruped animal's gaits and switch them according to the body joints' type and the slope angle [37]. Furthermore, even if the legs' number increased to six or more, a machine generates gaits [38]. These experiments suggest that even machines without a control mechanism can generate gaits by using gravity. The finding that walking machines produce different gait depending on body structure may be related to the fact that animals have different gait for different species. Realizing a robot that can actively walk requires studying the gait generation mechanism, including the actuator's driving method.

Recently, a quadruped robot system with joints using servomotors controlled by decoupled mathematical oscillators based on the active rotator model has been proposed [39–41]. The quadruped robot's legs are controlled according to the oscillator's phase individually. Feeding back each foot's pressure to the oscillators to accelerate and decelerate the joint's angular velocity has generated the phase difference (i.e., gait). The quadruped robot could generate an animal's gait according to the pressures. Another robot controlled in the same method could switch the gaits according to its moving speed [42]. These results suggest the effectiveness of using the difference in pressure on each foot to generate gaits. Although they did not design the oscillator they used to control the joint on a biological basis, the suggestion that the reaction force the leg receives from the floor is closely related to the gait is consistent with the results of physiological experiments [16, 17].

The authors speculate that spike firing has significant roles in information processing in the nervous system. We are studying robot control using pulse-type hardware neuron models (P-HNMs) that can output the spike firing (action potential) the same as a biological neuron [43]. Our research aims to develop a simple and efficient control method for robots using the artificial motor nervous system and central nervous systems. Hardware implementation will be advantageous in a large scale network system. The authors developed a quadruped robot system that implemented an active gait generation mechanism using P-HNMs. The mechanism is similar to the peripheral nervous system in that independent P-HNMs control each limb individually [44].

This chapter describes the active gait generation method for a quadruped robot. Firstly, the authors introduce the components of our quadruped robot system. Secondly, we will discuss the gait generation method, and finally, we will show the experimental result.

2. Quadruped robot system

Figure 1 shows our constructed quadruped robot system. This section describes the individual components of the quadruped robot system.

2.1 Mechanical and electrical components

The mechanical components of the robot system consist of the body frame and four-leg units. **Figure 2** shows the structure of the robot's body. We machined Part 1, 2, 3, and 4 from aluminum alloy sheets using the computerized numerical control (CNC) machining system and bender. Also, we formed Part A and B using a 3D printer. Part 2 and Part 3 were jointed together by Part 1 to form the body frame to connect the legs (**Figure 3**). The leg units consist of Part A, B, and two Part 4 and servomotors KRS-2552 (Kondo Kagaku co., ltd, available online at https://kondo-robot.com/ [45]). All the leg units have the same structure. The leg length is 138 mm (from joint axis to toe), the distance between the front and rear legs is 175 mm, and the distance between the left and right legs is 101 mm. The total weight of the robot system is approximately 1.1 kg. We gave the robot system two joints for each leg, the minimum number needed to walk. This robot system has no degrees of freedom except for legs.

The robot system's electrical components consist of self-inhibited P-HNM circuit boards, pressure sensors FSR402 (Interlink Electronics, Inc., available online at https://www.interlinkelectronics.com/ [46]), a single-board microcontroller Arduino DUE, and a peripheral circuit board. The pressure sensors have attached to the feet shown in **Figure 4**. Also, we mounted a battery and Bluetooth module to experiment without physical connections for power supply and data logging. The self-inhibited P-HNM and the peripheral circuit board are described later.

2.2 Self-inhibited P-HNM

The self-inhibited P-HNM consists of a cell body model and an inhibitory synaptic model. **Figure 5** shows the schematic and circuit diagram of the self-inhibited P-HNM. **Figure 5(a)** shows the connection between the cell body model and the inhibitory neuron model. A large circle represents the cell body model, and a black

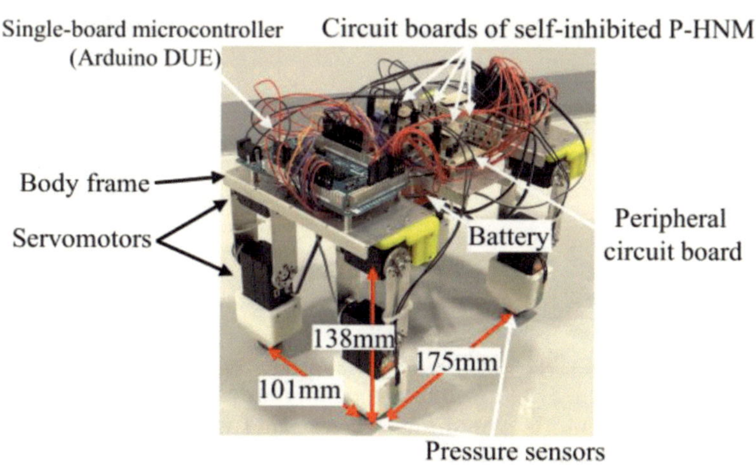

Figure 1.
Quadruped robot system.

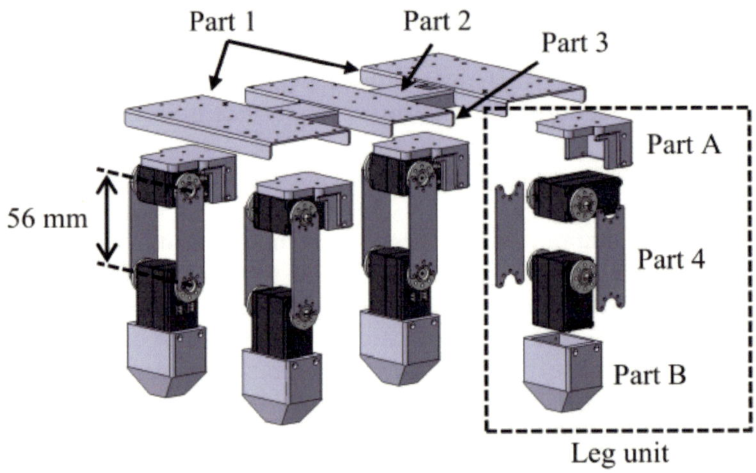

Figure 2.
Structure of robot's body. The quadruped robot system's body consists of a body frame and four leg units.

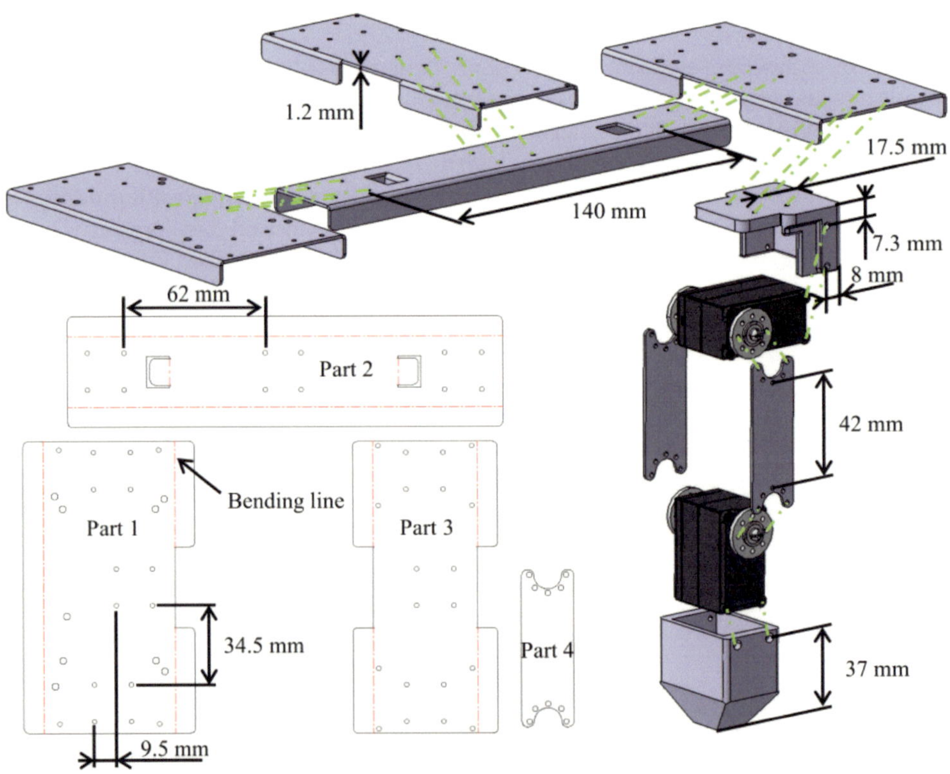

Figure 3.
Mechanical parts of the quadruped robot system. Part 1, 2, 3, and 4 were machined using the CNC system and bender.

circle and arc representing the inhibitory neuron model. The cell body model and the inhibitory synaptic model are mimicking several functions of a biological neuron. The cell body model includes a voltage control-type negative resistance, an equivalent inductance, resistors R_1 and R_2, and a membrane capacitor C_M. The voltage control-type negative resistance circuit with equivalent inductance consists of an n-channel MOSFET M_1, a p-channel MOSFET M_2, a voltage source V_A, a leak

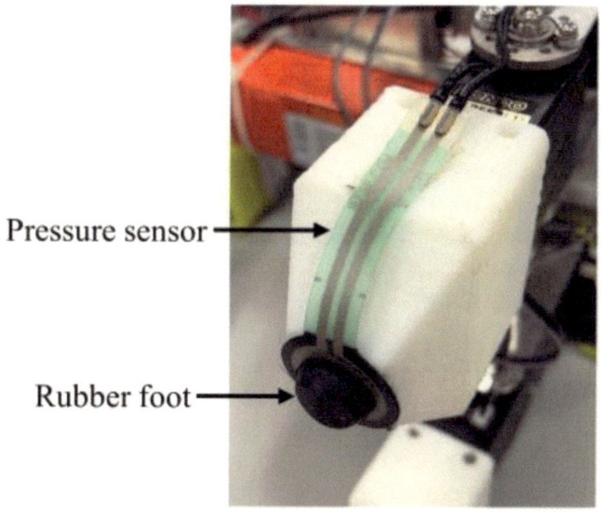

Figure 4.
Structure of each leg tip. The pressure sensor and rubber foot are attached to the end of Part B.

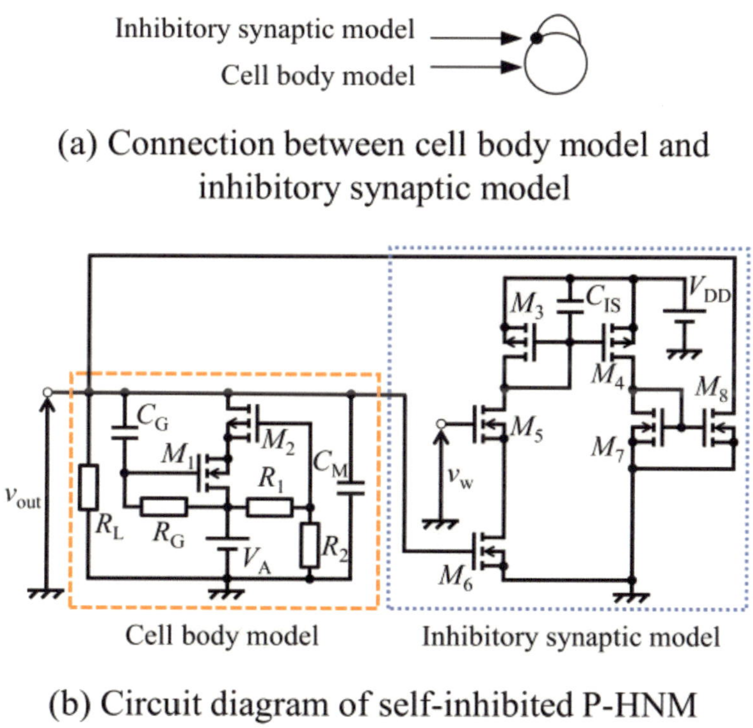

Figure 5.
Schematic diagram of self-inhibited P-HNM. According to the synaptic weight control voltage v_w*, the inhibitory synaptic model inhibits the cell body model's pulse generation.*

resistor R_L, another resistor R_G, and a capacitor C_G. The cell body model generates oscillating patterns of electrical activity v_{out}. More detail of the cell body model is described in [43]. The inhibitory synaptic model consists of simple current mirror circuits. The inhibitory synaptic model inhibits the cell body model's pulse generation by pulling out current from the cell body model. The strength of the inhibition can vary with synaptic weight control voltage v_w.

Figure 6 shows the simulation result of the waveform output by self-inhibited P-HNM. We changed v_w applied to the self-inhibited P-HNM in the middle of this simulation. The circuit constants are C_{IS} = 3.3 μF, C_G = 47 pF, C_M = 10 pF, R_1 = 15 kΩ, R_2 = 20 kΩ, R_G = 8.2 MΩ, R_L = 10 kΩ, $M_{1, 5, 6, 7, 8}$: BSS83, $M_{2, 3, 4}$: BSH205, V_{DD} = 5.0 V, V_A = 3.5 V. **Figure 7** shows the result of the measured relation between the pulse period T and the synaptic weight control voltage v_w applied to the self-inhibited P-HNM. The curve in **Figure 8** results from approximating the plotted points with a second-order polynomial, the region we used to control the robot.

2.3 Leg controlling system

We connected these components to realize the quadruped robot system with an active gait generating function. **Figure 8** shows the single-leg controlling system's schematic diagram, including the peripheral circuit's circuit diagram. The outline of the peripheral circuit board and self-inhibited P-HNM circuit board is described in **Figure 9**. The peripheral circuit consists of a low-pass filter, buffer, and voltage dividing circuits. The low-pass filter consists of a resistor R_F and a capacitor C_F. The buffers consist of operational amplifier U_1 and U_2. The voltage dividing circuits consist of R_{D1}, R_{D2}, and R_{D3}. The circuit constants of peripheral circuit are C_F = 3.3 μF,

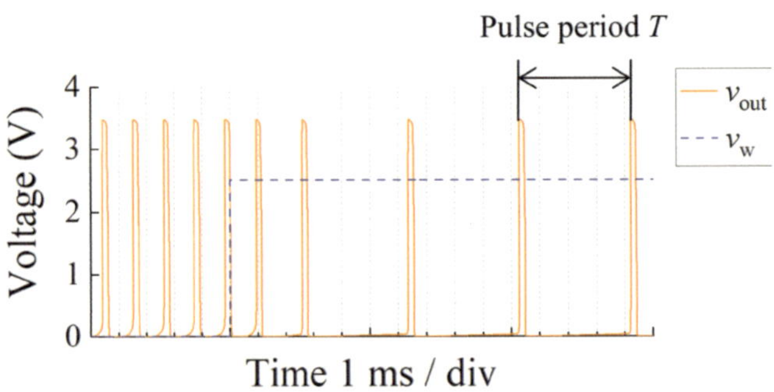

Figure 6.
Example of self-inhibited P-HNM's output waveform (simulation result). The P-HNM outputs pulses periodically. The pulse period increases with synaptic weight control voltage v_w.

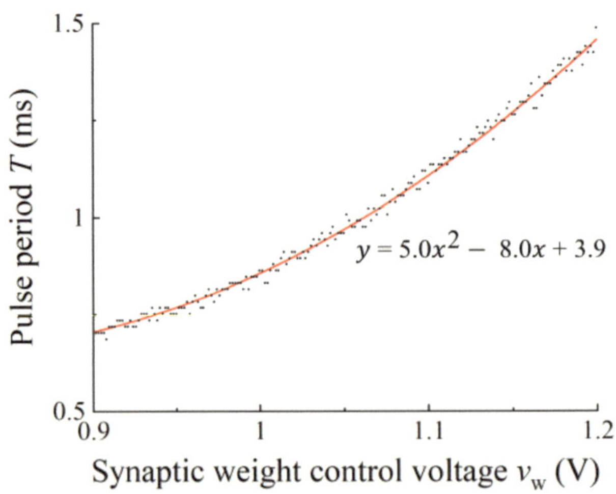

Figure 7.
Pulse period characteristics of self-inhibited P-HNM (experimental result). The range of v_w *we used to control the legs can be described in this equation.*

R_F = 11 kΩ, $R_{D1, D2, D3}$ = 11 kΩ, $U_{1, 2}$: LMC6032. We input the microcontroller's PWM output to the self-inhibited P-HNM circuit boards as an analog output through the peripheral circuits' low-pass filter. We connected the self-inhibited P-HNM's output voltage to the Arduino DUE's interrupt pin through the peripheral circuit.

We set two commands in the microcontroller to move the legs individually. One is to convert the reading from the pressure sensors into the synaptic weight control voltage v_w and input it to the self-inhibited P-HNM circuit boards. The other is changing the servomotors' angle by a constant angle each time the voltage input to the interrupt pin exceeds Arduino DUE's interrupt trigger voltage (approximately 1.7 V). We defined four-foot target points to create the foot trajectory shown in **Figure 10**. The foot passes through the target points and moves along the trajectory when the robot system changes the joint's angle. We set the microcontroller to process these commands individually for four legs. The overview of these commands is shown in **Figure 11**.

Each leg moves by a constant angle each time the self-inhibited P-HNM circuit boards output a pulse and the period at which the P-HNMs output a pulse varies

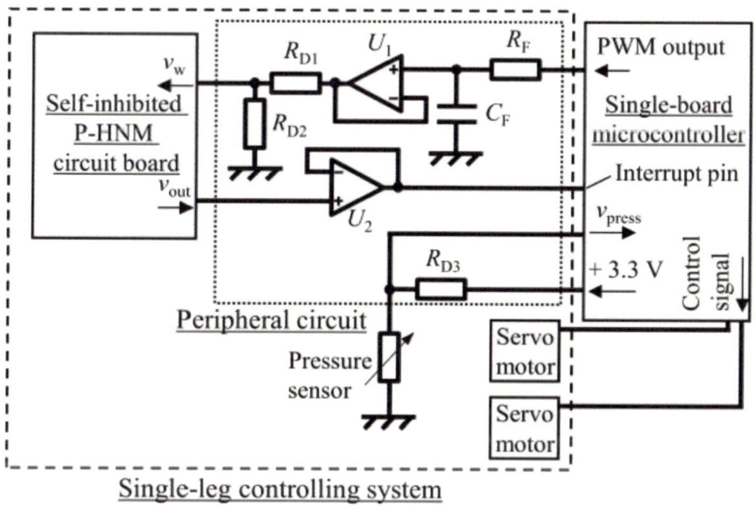

Figure 8.
Schematic diagram of the single-leg controlling system. We implemented peripheral circuits and the self-inhibited P-HNM circuit boards to the peripheral circuit board.

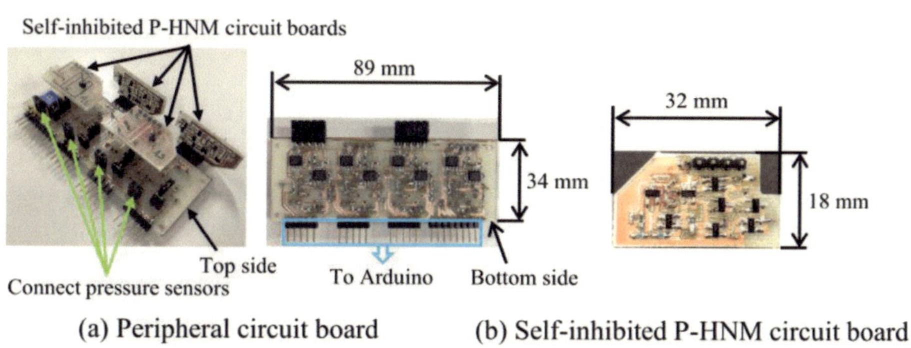

Figure 9.
Outline of the peripheral circuit board and self-inhibited P-HNM circuit board. (a) Peripheral circuit board and (b) self-inhibited P-HNM circuit board. Four self-inhibited P-HNM circuit boards and pressure sensors are connected to the peripheral circuit board.

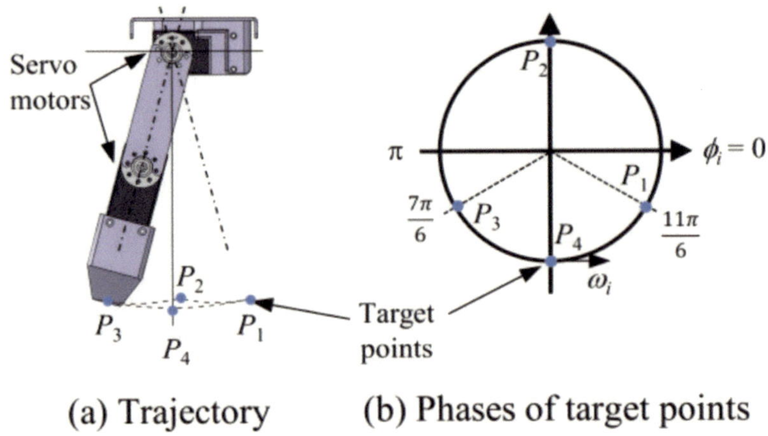

Figure 10.
Foot trajectory. The feet draw a trajectory through target points. (a) the trajectory and (b) target point's phases.

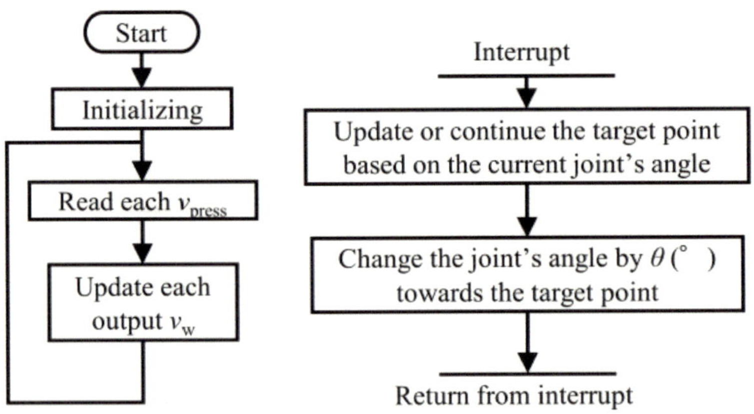

Figure 11.
Flow chart of the set commands. Interrupt command is executed when the v_{out} output by each self-inhibited P-HNM board exceeds the interruption trigger voltage.

depending on the pressure. Therefore, the speed at which each leg of the robot moves varies depending on the pressure on the feet.

3. Gait generation method

The following equations express the relation between the speed of moving legs and the pressure on the feet. The microcontroller controlled the legs individually. Therefore, some parameters are different for each leg. In the following equations, the subscript "*i*" means the parameter for the *i*th leg. The angular velocity of moving legs ω_i can be described as the following equation:

$$\omega_i = \frac{\theta}{T_i} \tag{1}$$

where θ is an actuation angle of servomotors each time the self-inhibited P-HNM circuit boards output a pulse. The synaptic weight control voltage v_w applied to the self-inhibited P-HNM circuit boards can be described as the following equation:

$$v_{wi} = \sigma\, v_{pressi} \tag{2}$$

where v_{pressi} is the applied voltage to the microcontroller depending on output by the pressure sensors. σ is a constant for converting v_{pressi} to v_{wi} and represents the effect of pressure. From the approximate formula in **Figure 7**, the pulse period T_i of the output voltage of self-inhibited P-HNMs v_{out} can be described as the following equation:

$$T_i = 5.0{v_{wi}}^2 - 8.0v_{wi} + 3.9. \tag{3}$$

From these equations, ω_i can describe as the following equation. This equation indicates that the pressure on the foot reduces the angular velocity of moving the leg.

$$\omega_i = \frac{\theta}{5.0{v_{wi}}^2 - 8.0v_{wi} + 3.9}. \tag{4}$$

4. Experiment result

We put the robot system on a flat floor and experimented under two conditions: the robot's walking speed is slow and fast. To change the robot's walking speed, we changed the legs' angular velocity by changing θ. However, we did not change parameters such as σ. We set the initial phase of each leg to $3\pi/2$ and let the legs to start moving at the same time.

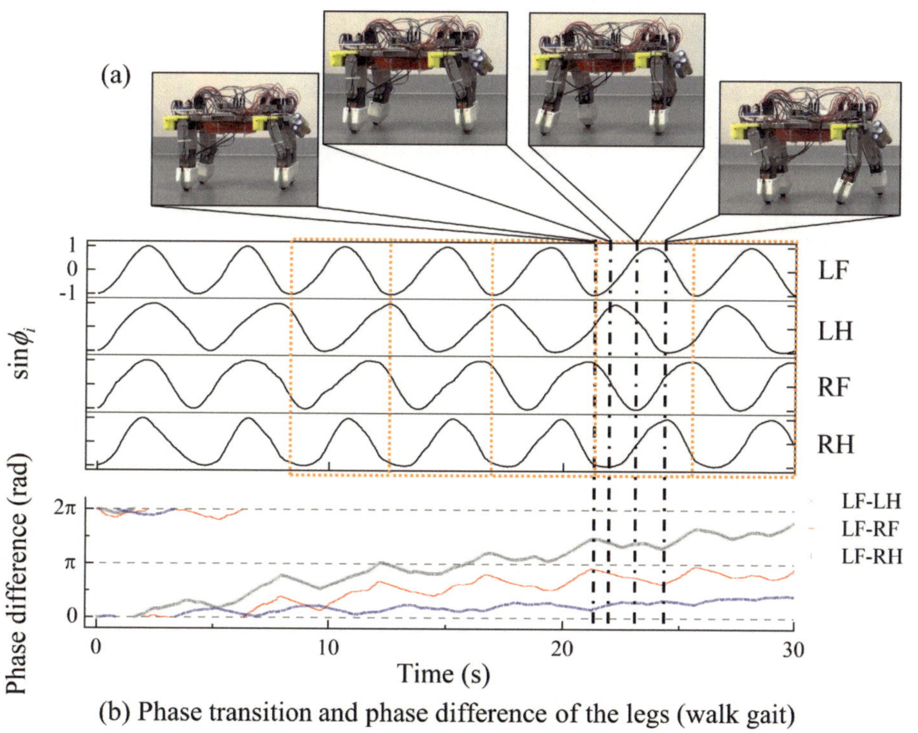

Figure 12.
Phase transition of the legs at low speed. (a) Walking quadruped robot system and (b) each leg's phase transition and phase difference from the LF. The robot system generated the walk gait from an upright position.

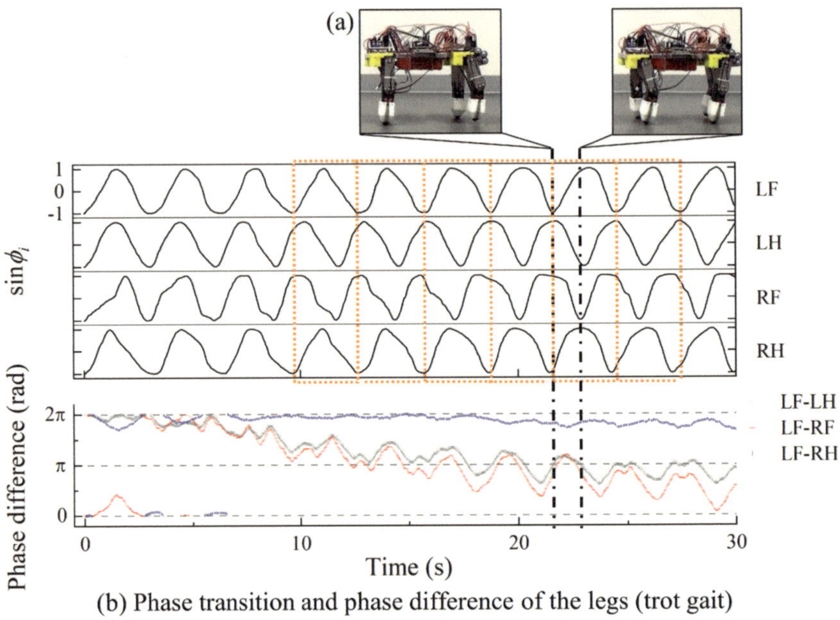

Figure 13.
Phase transition of the legs at high speed. (a) Trotting quadruped robot system and (b) each leg's phase transition and phase difference from the LF. The robot system generated the trot gait from an upright position.

Figure 12 shows the transition of each leg's phase ϕ_i. **Figure 12** also shows the phase difference between each leg and the left foreleg at low speed. The borders in **Figure 12** mean one cycle of gait. As **Figure 12** shows from the third step after the robot system start walking, each leg's phase differences were generated around 90° (0.5π rad). Also, the order of moving the legs is left fore (LF), right hind (RH), right fore (RF), and left hind (LH), which means that this gait is the same as the horse's walk gait. In this experiment, the legs' angular velocity while the legs were not on the floor was approximately 30°/s (0.52 rad/s). **Figure 13** shows the result at high speed. As **Figure 13** shows from the fourth step after the robot system start walking, each leg's phase difference was generated around 180° (π rad). Besides, the order of moving the legs is LF and RH, RF and LH, which means that this gait is the same as the horse's trot gait. In this experiment, the legs' angular velocity while the legs were not on the floor was approximately 51°/s (0.89 rad/s).

These results show that the quadruped robot system can generate gaits by reducing the legs' angular velocity depending on the pressure on the feet. Also, the robot system can generate different gaits depending on moving speed. Furthermore, the characteristics of the generated gaits are similar to the horse's gaits. In our control method, we confined the change factor in each leg's speed to feedback using weight-bearing balance. Therefore, we assume that the trigger for the break in the initial phase symmetry was slightly different in the robot's weight that the limbs were supporting. We have experimentally determined the parameters such as θ and σ that can stably produce these gestures. We expect that the dynamics simulator is necessary to determine these parameters quantitatively. In the future, we will use it to analyze in detail how the parameters affect the gaits.

5. Conclusions

In this chapter, the authors constructed a quadruped robot controlled by the active gait generating method individually for four legs. The method is simply

DOI: http://dx.doi.org/10.5772/intechopen.95760

reducing the robot's legs' moving speed according to the pressures of feet. We implemented pulse-type hardware neuron models (P-HNMs) to the method. We conducted experiments under two conditions: when the robot's walking speed is slow and fast. As a result, the robot system actively generated phase differences of each leg. By analyzing the experimental results, we clarified the process of gait generation. Also, we confirmed that the generated phase differences were similar to the horse's gaits of walk and trot. These results suggest that quadruped robots can spontaneously generate gaits according to the environment by our proposed mechanism. Furthermore, it shows that animals may generate gait using a similarly simple method because P-HNM mimics biological neurons' function. In the future, we intend to use a kinetic simulator to determine the basis of how the robotic system generates gait.

Video materials

Additional video materials available at: https://bit.ly/3i91LbI

Acknowledgements

This work was supported by Nihon University Multidisciplinary Research Grant for (2020) and supported by Research Institute of Science and Technology Nihon University College of Science and Technology Leading Research Promotion Grant. Also, the part of the work was supported by JSPS KAKENHI Grant Number JP18K04060. The authors appreciated the Nihon University Robotics Society (NUROS).

Author details

Yuki Takei, Katsuyuki Morishita, Riku Tazawa and Ken Saito*
Nihon University, Chiba, Japan

*Address all correspondence to: kensaito@eme.cst.nihon-u.ac.jp

References

[1] Raibert M, Blankespoor K, Nelson G, Playter R. BigDog, the Rough-Terrain Quadruped Robot. In IFAC Proceedings; 2008. p. 10822-10825

[2] Ruina D, Chenxing J, Ning L, Zhenjie L, Honglei Z, Bo S. The analysis of key technologies for advanced intelligent quadruped robots. In International Conference on Control and Cybernetics; 2019; Tokyo. Japan; 2019. p. 70-74

[3] Habib MK, Watanabe K, Izumi K. Biomimetics robots from bio-inspiration to implementation. In: Proceedings of the 33rd annual conference of the IEEE Industrial Electronics Society; 5-8 November 2007; Taipei. Taiwan: IEEE; 2008. pp. 143-148

[4] Habib MK: Biomimetics: innovations and robotics. International Journal of Mechatronics and Manufacturing Systems (IJMMS). 2011;4(2):113-134. DOI: 10.1504/IJMMS.2011.039263

[5] Marder E, Bucher D: Central pattern generators and the control of rhythmic movements. Current Biology. 2001; 11:R986-R996. DOI: 10.1016/S0960-9822(01)00581-4

[6] Selverston AI, Ayers J: Oscillations and oscillatory behavior in small neural circuits. Biol Cybern. 2006;95:537. DOI: 10.1007/s00422-006-0125-1

[7] McMahon TA: The role of compliance in mammalian running gaits. J Exp Biol. 1985;115:263-282. DOI: 10.1002/jez.a.334

[8] Yamanobe A, Hiraga A, Kubo K: Relationships between stride frequency, stride length, step length and velocity with asymmetric gaits in the thoroughbred horse. Japanese Journal of Equine Science. 1992;3:143-148. DOI: 10.1294/jes1990.3.143

[9] Bhatti Z, Waqas A, Mahesar W, Karbasi M: Gait analysis and biomechanics of quadruped motion for procedural animation and robotic simulation. Bahria University Journal of Information & Communication Technologies. 2017;10:2.

[10] Coulmance M, Gahéry Y, Massion J, Swett JE: The placing reaction in the standing cat: A model for the study of posture and movement. Experimental Brain Research. 1979;37:265-281. DOI: 10.1007/BF00237713

[11] Duysens J, Pearson KG: Inhibition of flexor burst generation by loading ankle extensor muscles in walking cats. Brain Research. 1980;187(2):321-332. DOI: 10.1016/0006-8993(80)90206-1

[12] Hoyt D, Taylor C: Gait and the energetics of locomotion in horses. Nature. 1981;292:239-240. DOI: 10.1038/292239a0

[13] Taylor CR: Force development during sustained locomotion: a determinant of gait, speed and metabolic power. Journal of Experimental Biology. 1985;115:253-262.

[14] Wisleder D, Zernicke RF, Smith JL: Speed-related changes in hindlimb intersegmental dynamics during the swing phase of cat locomotion. Exp Brain Res. 1990;79(3)]651-660. DOI: 10.1007/BF00229333

[15] Vilensky JA, Libii JN, Moore AM: Trot-gallop gait transitions in quadrupeds. Physiology & Behavior. 1991;50:835-842. DOI: 10.1016/0031-9384(91)90026-K

[16] Cartmill M, Lemelin P, Schmitt D: Support polygons and symmetrical gaits in mammals. 2002;136:401-420. DOI: 10.1046/j.1096-3642.2002.00038.x

[17] Carlo M, Biancardi, Alberto E, Minetti: Biomechanical determinants of transverse and rotary gallop in cursorial mammals. Journal of Experimental Biology. 2012;215:4144-4156. DOI: 10.1242/jeb.073031

[18] Sarah J, Hilary M, Clayton M: Sagittal plane ground reaction forces, centre of pressure and centre of mass in trotting horses. The Veterinary Journal. 2013;e14-e19. DOI: 10.1016/j.tvjl.2013.09.027

[19] Grillner S: Locomotion in vertebrates: central mechanisms and reflex interaction. Physiological Reviews. 1975;55(2):247-304. DOI: 10.1152/physrev.1975.55.2.247

[20] Orsal D, Cabelguen JM, Perret C: Interlimb coordination during fictive locomotion in the thalamic cat. Exp Brain Res. 1990;82:536-546. DOI: 10.1007/BF00228795

[21] Cruse H, Warnecke H: Coordination of the legs of a slow-walking cat. Exp Brain Res. 1991;89:147-156. DOI: 10.1007/BF00229012

[22] Bertram JEA: Gait as solution, but what is the problem? Exploring cost, economy and compromise in locomotion. The Veterinary Journal. 2013;198:e3-e8. DOI: 10.1016/j.tvjl.2013.09.025

[23] Grillner S, Zangger: On the central generation of locomotion in the low spinal cat. Experimental Brain Research. 1979;34:241-261. DOI: 10.1007/BF00235671

[24] Frigon A, Rossignol S: Experiments and models of sensorimotor interaction during locomotion. Biological Cybernetics. 2006;95:607-627. DOI: 10.1007/s00422-006-0129-x

[25] Bellardita C, Kiehn O: Phenotypic characterization of speed-associated gait changes in mice reveals modular organization of locomotor networks. Current Biology. 2015;25(11):1426-1436. DOI: 10.1016/j.cub.2015.04.005

[26] Ruber L, Takeoka A, Arber S: Long-distance descending spinal neurons ensure quadrupedal locomotor stability. Neuron. 2016;92(5):1063-1078. DOI: 10.1016/j.neuron.2016.10.032

[27] Delcomyn F: Neural basis of rhythmic behavior in animals. Science. 1980;210:492-498. DOI: 10.1126/science.7423199

[28] Arshavdky YI, Deliagina TG, Orlovsly GN: Central pattern generators: mechanisms of operation and their role in controlling automatic movements. Neuroscience and Behavioral Physiology. 2016;46(6):696-718. DOI: 10.1007/s11055-016-0299-5

[29] Yuasa H, Ito M: Coordination of many oscillators and generation of locomotory patterns. Biological Cybernetics. 1990;63:177-184. DOI: 10.1007/BF00195856

[30] Ito S, Yuasa H, Luo ZW, Ito M, Yanagihara D: A mathematical model of adaptive behavior in quadruped locomotion. Biological Cybernetics. 1998;78:337-347. DOI: 10.1007/s004220050438

[31] Kukikkaya R, Proctor J, Holmes P: Neuromechanical models for insect locomotion: Stability, maneuverability, and proprioceptive feedback. Chaos: An Interdisciplinary Journal of Nonlinear Science. 2009;19. DOI: 10.1063/1.3141306

[32] Ishii T, Masakado S, Ishii K. Locomotion of a quadruped robot using CPG. In IEEE International Joint Conference on Neural Networks; 25-29 July 2004; Budapest. Hungary: IEEE; 2005. p. 3179-3184

[33] Li X, Wang W, Yi J. Foot contact force of walk gait for a quadruped robot.

In IEEE International Conference on Mechatronics and Automation; 7-10 August 2016; Harbin. China: IEEE; 2016. p. 659-664

[34] Liu H, Jia W, Bi L. Hopf oscillator based adaptive locomotion control for a bionic quadruped robot. In International Conference on Mechatronics and Automation; 6-9 August 2017; Takamatsu. Japan: IEEE; 2017. p. 949-954

[35] Habu Y, Yamada Y, Fukui S, Fukuoka Y. A simple rule for quadrupedal gait transition proposed by a simulated muscle-driven quadruped model with two-level CPGs. In IEEE International Conference on Robotics and Biomimetics; 12-15 December 2018; Kuala Lumpur. Malaysia: IEEE; 2018. p. 2075-2081

[36] McGeer T: Passive Dynamic Walking. The International Journal of Robotics Research. 1990;9(2):62-82. DOI:10.1177/027836499000900206

[37] Nakatani K, Sugimoto Y, Osuka K: Demonstration and Analysis of Quadrupedal Passive Dynamic Walking, Advanced Robotics. 2009;23:483-501. DOI: 10.1163/156855309X420039

[38] Sugimoto Y, Yoshioka H, Osuka K: Development of Super-multi-legged Passive Dynamic Walking robot "Jenkka-III". SICE Annual Conference 2011; Tokyo. IEEE;2011. p.576-579

[39] Owaki D, Kano T, Nagasawa K, Tero A, Ishiguro A: Simple robot suggests physical interlimb communication is essential for quadruped walking. J R Soc Interface. 2013;10(78):20120669. DOI: 10.1098/rsif.2012.0669

[40] Shinomoto S, Kuramoto Y: Phase Transitions in active rotator systems. Progress of Theoretical Physics. 1986;75:1105-1110. DOI: 10.1143/PTP.75.1105

[41] Shinomoto S, Kuramoto Y: Cooperative Phenomena in Two-Dimensional Active Rotator Systems. 1986;75(6):1319-1327. DOI: 10.1143/PTP.75.1319

[42] Owaki D, Ishiguro A: A quadruped robot exhibiting spontaneous gait transitions from walking to trotting to galloping. Sci Rep 7;2017:277 10.1038/s41598-017-00348-9

[43] Saito K, Ohara M, Abe M, Kaneko M, Uchikoba F: Gait generation of multilegged robots by using hardware artificial neural networks. Advanced Applications for Artificial Neural Networks. IntechOpen. 2017. DOI: 10.5772/intechopen.70693

[44] Takei Y, Morishita K, Tazawa R, Katsuya K, Saito K: Non-programmed gait generation of quadruped robot using pulse-type hardware neuron models. Artif Life Robotics. 2020. DOI: 10.1007/s10015-020-00637-z

[45] Kondo Kagaku co., ltd. [Internet]. Available from: https://kondo-robot.com/ [Accessed: 2020-11-14]

[46] Interlink Electronics, Inc. [Internet]. Available from: https://www.interlinkelectronics.com/ [Accessed: 2020-11-14]

Chapter 3

From Insect Vision to a Novel Bio-Inspired Algorithm for Image Denoising

Manfred Hartbauer

Abstract

Night active insects inspired the development of image enhancement methods that uncover the information contained in dim images or movies. Here, I describe a novel bionic night vision (NV) algorithm that operates in the spatial domain to remove noise from static images. The parameters of this NV algorithm can be automatically derived from global image statistics and a primitive type of noise estimate. In a first step, luminance values were ln-transformed, and then adaptive local means' calculations were executed to remove the remaining noise without degrading fine image details and object contours. Its performance is comparable with several popular denoising methods and can be applied to grey-scale and color images. This novel algorithm can be executed in parallel at the level of pixels on programmable hardware.

Keywords: night vision, spatial integration, contrast enhancement, noise reduction, denoising, image enhancement, image processing, local means calculation

1. Introduction

Some insect species have attracted the attention of researchers due to their astonishing visual abilities under extremely dim light conditions [1–3]. These insects cope with noise that degrades visual information and has multiple origins: the sparsity of photons results in shot noise, which is overlaid by transducer noise. To increase the sensitivity of compound eyes, the clusters of photoreceptor cells direct light on to a certain part of the associated rhabdom in order to gather the photons from a wide field of view. In other, nocturnal insect species (e.g., *Megalopta genalis*) neurons in the brain sum up the information provided by individual ommatidia forming the apposition eye. Insects equipped with such neural apposition eyes can even see at star light conditions [3]. Filtering in the spatial [4] and temporal domains is mirrored in some denoising algorithms that are available for cleaning up noisy films (e.g., [5]). However, there is still a lack of image enhancement methods that improve the quality of underexposed static images while avoiding artifacts and preserving image sharpness. The elimination of noise from static images is usually a challenging task for any denoising algorithm, because the temporal domain is not available for filtering.

The quality of images taken under dim light conditions is also often reduced by imperfections in the sensor itself ('sensor grain noise') and shot noise. Generally, dim images have a very limited luminance range, which limits the content of available information. If any measure is undertaken to improve image contrast,

IntechOpen

such as the traditional method of histogram stretching, sensor noise is unavoidably amplified. Therefore, the goal of image enhancement is to preserve as many image details as possible while eliminating noise. Typically denoising can be achieved by the application of linear and nonlinear filters. Linear filters take the forms of smoothing or low-pass, sharpening, Laplacian, un-sharp masking, or high-boost filters. Nonlinear filters include order statistic filters such as minimum, median, and maximum filters (for a review of methods see [6, 7]).

Simple denoising techniques, such as linear smoothing or median filtering, can reduce noise, but at the same time smooth away edges, so that the resulting image becomes blurry. A popular alternative denoising method is total variation (TV) denoising, which has been described by Rudin et al. [8]. This method minimizes the total variation of the luminance values that can mainly be attributed to noise. The TV regularization method preserves salient edges while effectively removing noise. Lee et al. [9] published a framework for a Moving Least Squares Method with Total Variation Minimizing Regularization and Yoon et al. [10] improved the preservation of fine image details by developing an adaptive, TV-minimization-based image enhancement method (ATVM). Bilateral filtering described by Tomasi and Manduchi [11] is another powerful non-linear denoising algorithm that preserves object contours. Here, denoising is based on the spatial distance of surrounding pixels relative to an output pixel and its grey value difference. Bilateral filtering is fast but the tuning of parameters is rather difficult (see Zhang and Gunturk [12]) and staircase effects and inverse contours are known possible artifacts. Another possibility is the operations that are performed on a Fourier transform of the image rather than on the image itself. The techniques that fall under this category include low pass, high pass, homomorphic, linear and root filtering. Fourier-transformed images are filtered and inverse transformed to reduce noise and prevent blurring effects. The disadvantages of frequency domain methods are that they introduce certain artifacts and cannot simultaneously enhance all parts of the image very well. In addition, it is difficult to automate the image enhancement procedure. Despite these drawbacks, frequency filtering of similarity maps has proved to be a powerful method for image denoising (BM3D, published by Maggioni et al. [13]; see original work of Dabov et al. [14]). This method divides the image into small pieces (2D blocks) and, after 3D transformation of similar blocks, the filtering process eliminates noise while leaving the object details mostly untouched. In addition, wavelet-domain hidden Markov models have been applied to image denoising with fascinating results, especially when applied to diagnostic images [15–17].

In order to reduce the computational time required by complex image-processing algorithms such as edge detectors, homomorphic filtering, and image segmentation, general-purpose computing methods using graphics processing units were developed [18, 19]. More simple, computationally less demanding algorithms were developed as another strategy to reduce processing time. For example, the simple piecewise linear (PWL) function sharpens image edges and reduces noise simply by evaluating the luminance of pixels in a window of 3 × 3 pixels around each pixel [20]. Its effects can easily be controlled by varying only two parameters. Such simple algorithms can be implemented in reconfigurable hardware in the form of field-programmable gate arrays (FPGA), which is considered a practical way to obtain high performance when using computationally-intensive image processing algorithms [21, 22]. Performing parallel operations on hardware significantly reduces processing time, but simple algorithms are easier to implement on programmable hardware compared to mathematically complex ones.

Here, I describe a rather simple, bio-inspired algorithm that can be used to enhance the contrast of dim images and remove noise without affecting fine image details much. It operates in the spatial domain at the level of pixels and can be run in parallel on FPGA hardware.

2. Bionic method of image denoising

2.1 Method overview

This novel night vision (NV) image enhancement method increases the quality of underexposed pictures by combining three subsequent image processing steps (see **Figure 1**). These are executed at the level of pixels which perform simple calculations to mimic the amplification of the transduction process in photoreceptors and the spatial integration of image information as known from nocturnal insects [2]. The photoreceptors of *Megalopta genalis* (Halictidae), a nocturnal bee found in the Neotropics, has a rather great gain of transduction, which decreases the signal-to-noise ratio and information capacity in dim light for an increased sensitivity. This amplification of visual information is mirrored in the first image processing step of this night vision method by performing a logarithmic transformation of pixel grey values (luminance values). Logarithmic transformation of luminance values leads to an increase of small values while high values remain rather constant. Therefore, image details of dark image regions become visible.

The photoreceptors of nocturnal insects generate slow and noisy visual signals that are spatially summed by second-order monopolar cells in the lamina [1]. Summing visual information from a wide angle of view leads to the reduction of noise and, thus, improves signal-to-noise ratio. The neuronal correlate for this can be found in large dendritic trees of lamina interneurons in *M. genalis*. However, to prevent image blur, spatial summation should be small in image areas where contrast is high and large in more homogeneous image regions. This "adaptive spatial averaging"

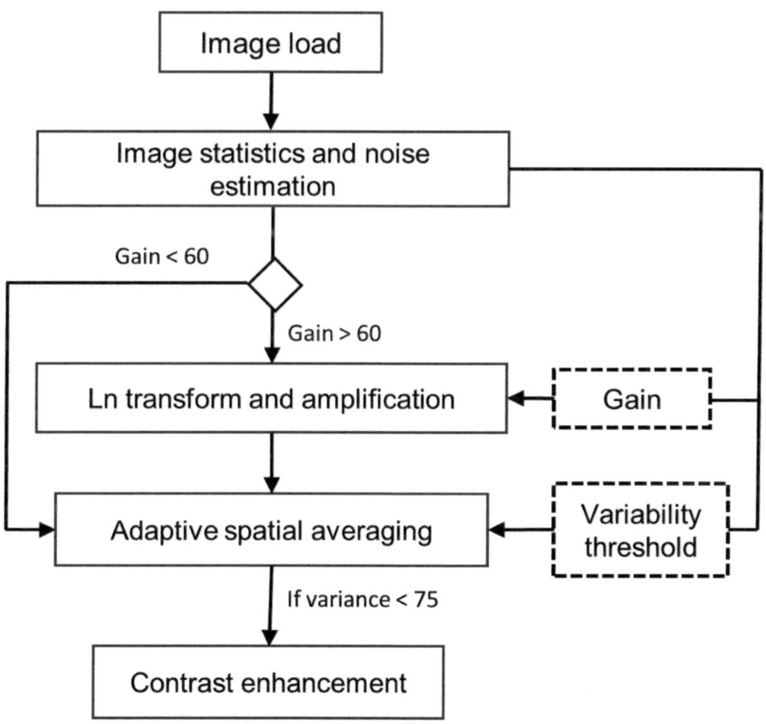

Figure 1.
Schema of image-processing steps that make up the NV algorithm. Global image statistics and a simple method of noise estimate were used to derive the parameters "gain" and "variability threshold". These were used in subsequent image-processing steps to enhance the quality of dim images and remove noise. Images with average brightness skip "Ln transformation" and the "contrast enhancement" routine was only applied to images that exhibited low variance among their luminance values.

is performed in the second image processing step of this night vision method and preserves object contours and image sharpness. This procedure assumes a higher variability of luminance values near object contours as compared to homogeneous image regions. Thus, circles in which local luminance values can be averaged may not exceed a critical variability of grey values (*threshold_var*). Adaptive averaging is performed at the level of pixels and evaluates the variability of local grey values to find the dimension of a circle in which the variability of grey values remains below the predefined variability threshold. After exceeding this threshold, the average of grey values of pixels belonging to this circle is calculated and stored at the central pixel. As a final processing step an automatic contrast-enhancement procedure was applied by means of linear histogram stretching. Two parameters (gain and variability threshold) are essential for this method and were derived from global image statistics and a simple kind of noise estimate. The image-enhancement algorithm described here was developed using Netlogo 5.2 (developed by Uri Wilensky; http://ccl.northwestern.edu/netlogo/), a multi-agent programming environment that allows the parallel execution of commands at the level of pixels (named patches in the Netlogo language).

2.2 Import pictures

Dim and noisy images were imported into Netlogo using the command "import-pcolors", which transformed the luminance values of pixels into the grey values of patches. This Netlogo function limits the luminance values of grey-scale images to the range between 0 (black) to 9.9 (white), whereby the total number of possible grey values is 110. Although image information is reduced by this function, the human eye is unable to recognize any difference between images having a grey value range of 110 or higher.

2.3 Image statistics

For the automatic adjustment of the two parameters gain and variability threshold, a simple calculation of the global image statistics was performed and image noise was estimated. The median and variance of the grey values of all pixels were analyzed and saved as *median_grey* and *var_grey*, respectively. Before processing 'rgb' images, the parameter *median-grey* was calculated by averaging over the luminance values of all color channels. In contrast, *var_grey* was calculated from the channel with the highest average luminance value. The parameter *gain* acts as a factor of ln-transformed grey values (adjusts image brightness) and was derived from *median_grey* according to Eq. 1. *gain* cannot be smaller than 1.

$$gain = 5 - median\_grey \tag{1}$$

2.4 Noise estimation

A rather simple estimation of noise can be obtained by summing up the differences between the grey values of each pixel and the average grey values of surrounding pixels in rather homogeneous regions of the image. For this purpose, the local average luminance value (*mean_grey_local*) and local variance of luminance values (*var_grey*) of surrounding patches were calculated in a circle with a radius of 4 pixels. Since brighter images tend to show a higher variability of grey values, the noise estimation was restricted to those pixels X_{NE} having circles in which *mean_grey_local* was smaller than the ratio between *var_grey* and (*median_grey* – 1) (see Eq. (2)). Noise estimation was computed according to Eq. (3), whereby X denotes the number of patches in the homogeneous image regions.

$$X_{NE} \mid mean\_grey\_local < var\_grey / (median\_grey - 1) \tag{2}$$

$$noise\_estimate = \sum_{N=1}^{X} abs(mean\_grey\_local - luminance) / X \tag{3}$$

The noise estimation of color images was restricted to the color channel with the highest average luminance value.

2.5 Parameter estimation

This NV algorithm derives all its parameters from global image statistics (*median_grey* and *var_grey*) and the *noise_estimate*. Eq. (4) defines the threshold that was used for the adaptive spatial averaging procedure. It was found empirically by testing numerous terms that allow to predict the parameters that result in high quality output images with respect to the conservation of image details and the (Peak signal-to-noise ratio) PSNR value. Eq. (4) was derived from manually-adjusted parameter combinations that were obtained from 10 different images exhibiting various noise levels and brightness values.

$$threshold\_var = 0.0002 * [(noise\_estimate + var\_grey) * median\_grey] + 0.92 + (gain/10) \tag{4}$$

If *noise_estimate* was smaller than 0.01, *threshold_var* was always set to 0.01, which is low enough to preserve fine image details, but high enough to remove the remaining noise.

2.6 Image processing step 1: logarithmic transformation

The logarithmic transformation of luminance values results in disproportionate amplification of the small grey values. The image contrast of very dim images was improved by calculating the natural logarithm of the luminance values of patches of each color channel and multiplication of the result by the *gain* factor using Eq. (5). Adding the constant 1.5 to the luminance values prevents them from becoming smaller than zero after ln-transformation. The result of this logarithmic transformation was stored as *greyLn* at the focal pixel.

$$greyLn = \ln(grey\_value + 1.5) * gain \tag{5}$$

2.7 Image processing step 2: adaptive spatial averaging

Image noise was widely removed by means of 'adaptive spatial averaging', a procedure that is executed by each pixel and evaluates the local variability of the *greyLn* values to calculate the radius of the circle in which grey value averaging is executed (locals means computation). Adaptive spatial averaging was executed in parallel at the level of pixels. Each pixel expands a circle in steps of one patch as long as the variability of the average *greyLn* values of the pixels of each color channel within this circle remained below *threshold_var*. Once the dimension of this circle was found, the *greyLn* values were averaged and saved as *grey_avg* value for the focal pixel. This averaging of *greyLn* values in the circle was calculated for each color channel separately. The maximum radius of the expanding circle was restricted to 10 pixels and the minimum size to 1 pixel.

2.8 Image processing step 3: enhancement of image contrast

After adaptive spatial averaging, image contrast can be enhanced if the grey value variability of the resulting image is low (variance <75). This was done by means of linear histogram stretching, which uses the lowest and highest *grey_avg* values of the image to calculate the resulting grey value of each patch. This was achieved by using Eq. (6), which sets the lowest value to 0 (black) and the highest value to 9.9 (white) while intermediate values are assigned to shades of grey.

$$grey\_value = \left\{ \left[grey\_avg - min\,(grey\_avg)\right] / \left[max\,(grey\_avg) - min\,(grey\_avg)\right] \right\} * 9.9 \tag{6}$$

2.9 Evaluation of performance

The performance of this algorithm was evaluated by calculating the peak signal-to-noise-ratio (PSNR) using the method described by Russo [20]. The result is given in dB and quantifies the difference between the noisy input image and the processed image. Higher dB values indicate better denoising performance. To evaluate PSNR between the input and the output images, images were exported from Netlogo in 'png' format and a Python script was programmed to execute the function "compare_psnr" offered in the "skimage.measure" library.

3. Results image denoising

The sequence of image-processing steps illustrated in **Figure 1** allowed the strong enhancement of the image contrast in dim images and simultaneously removed sensor noise. The performance of this simple, dynamic spatial domain-filtering algorithm depends on two parameters that can be estimated by evaluating global image statistics and executing a simple noise estimation method.

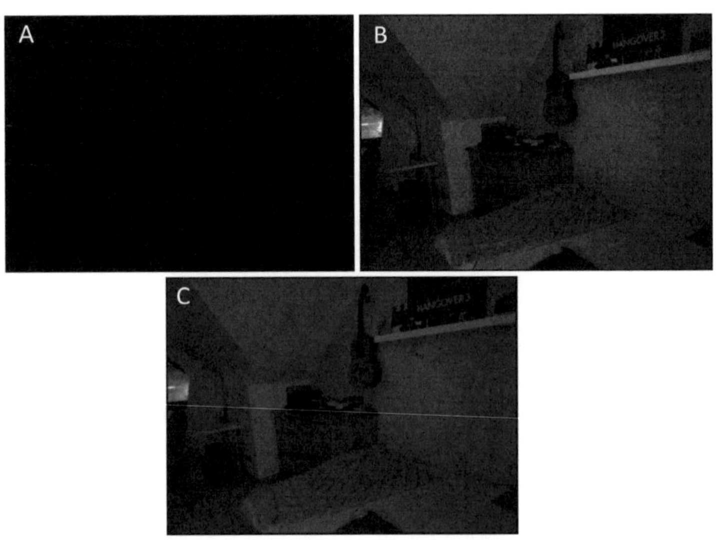

Figure 2.
Performance of the NV algorithm using a dim image of a mobile phone camera as input. (A) The original image is black. (B) After automatic histogram equalization sensor noise was amplified. (C) The NV-processed image exhibits a high level of image contrast and contains a low amount of noise. PSNR of C: 25.9 dB. Hartbauer holds the copyright of this picture.

Figure 3.
Performance of the NV algorithm using a dim natural image as input. (A) Noisy image of a village on Mallorca (printed with permission from Buades; see Buades et al. [6]). (B) Noise was amplified after histogram stretching. (C) The output of the NV algorithm yielded a reduced amount of noise (PSNR = 35.3 dB) and shows a higher quality regarding image details and contrast.

Application of this NV algorithm to an underexposed image that was taken in a very dim room resulted in an output image with a high level of contrast and rather low noise level (see **Figure 2C**). On the contrary, the automatic adjustment of brightness and contrast offered by commonly used image processing software either produced a rather dark image (similar as **Figure 2A**) or a contrasted image in which sensor noise was greatly enhanced (**Figure 2B**). The adaptive spatial averaging procedure preserved fine image details and object contours, while most noise was removed. This is mirrored in the high PSNR value of 25.9 dB when **Figure 2B** was compared with **Figure 2C**.

The NV algorithm enhanced image contrast and removed noise from the natural image showing a village (compare **Figure 3A** and **C**; PSNR = 35.3 dB), whereas automatic contrast enhancement amplified image noise greatly (**Figure 3B**). In comparison, denoising performed with an improved non local mean algorithm published by Buades et al. [6] removed noise slightly more effective (PSNR = 41.1 dB), but did not improve image contrast.

When Gaussian distributed noise with a standard deviation of 12.75 was added to an image of a bird (**Figure 4A**), the NV algorithm removed noise while retaining many fine image details (**Figure 4B**). This performance is reflected in a high PSNR value of 32.1 dB. The same noisy picture processed with a method that is based on the moving least squares (MLS) algorithm described by Lee et al. [9] resulted in the production of a rather blurry image.

The noisy grey-scale image, shown in **Figure 4C**, contains additive Gaussian noise with a standard deviation of 0.6. The output image of the NV algorithm

Figure 4.
Denoising performance of the NV algorithm using grey scale images as input. (A) The image of a bird (from Lee et al. [9]) was degraded by Gaussian distributed noise with σ = 12.75. (B) The NV algorithm removed noise (PSNR = 32.1 dB) and preserved many fine image details. (C) Additive Gaussian noise with σ = 0.6 was added to an image of an alarm clock. The NV algorithm removed noise (PSNR = 27.4 dB) without degrading fine image details (D). The picture was printed with permission from Springer Inc. The picture shown in (C) was printed with permission from Yoon et al. [10] (Creative Commons License).

(**Figure 4D**) contains many fine image details and the noise level was reduced by 27.2 dB (PSNR). This denoising performance is very similar to the adaptive total variation minimization-based image enhancement method (ATVM) described by Yoon et al. [10], which is mirrored in a similar PSNR value of 30.5 dB.

Adaptive spatial averaging as described here can also be used to remove noise from color images (see **Figure 5**). This NV algorithm successfully removed noise from the extremely noisy image showing a candle (**Figure 5A**). Denoising of this image resulted in a PSNR of 32.0 dB (**Figure 5B**), which is similar to the denoising performance of the GSM wavelet denoising described by Portilla et al. [23]. The NV algorithm also removed most noise from the image showing the face of an ostrich in **Figure 5C**, resulting in a PSNR value of 32.7 dB. Interestingly, the resulting image is rather sharp and still contains many fine image details, such as the hairs. In contrast, the novel image denoising method described by Liu et al. [4] created

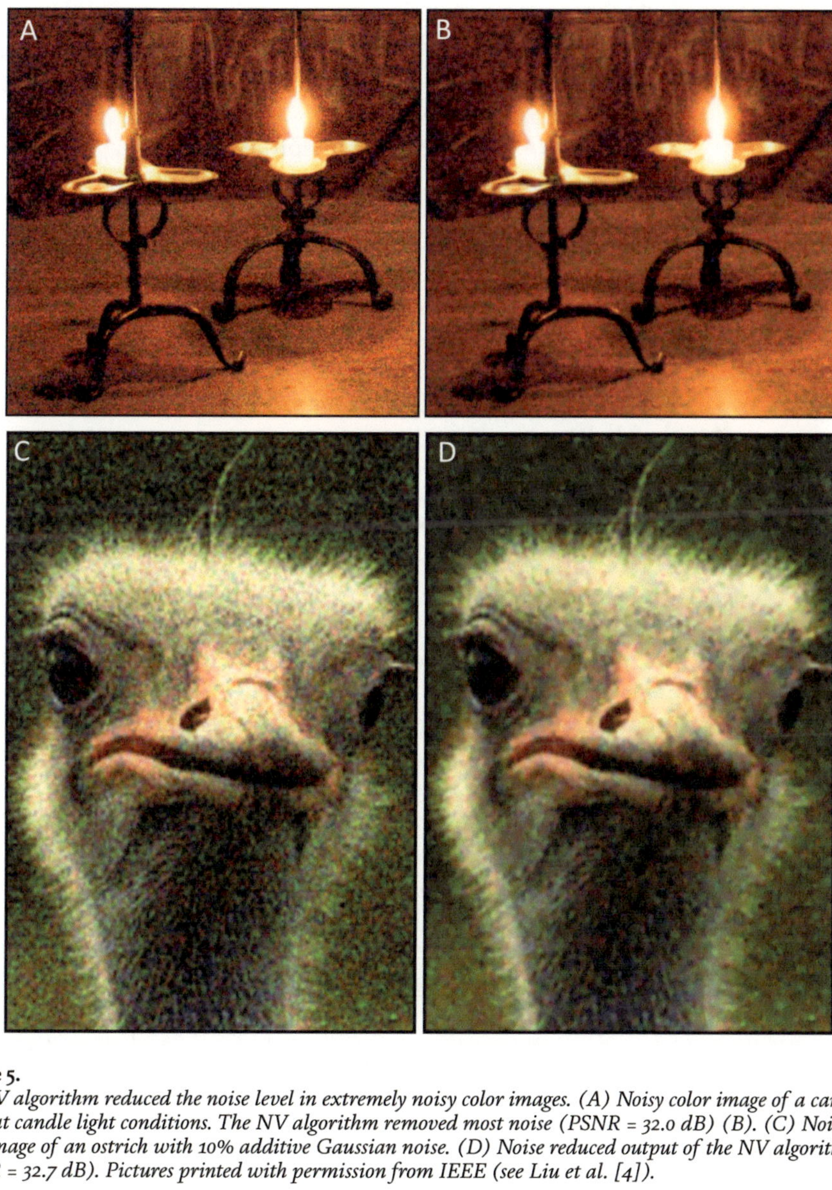

Figure 5.
The NV algorithm reduced the noise level in extremely noisy color images. (A) Noisy color image of a camera taken at candle light conditions. The NV algorithm removed most noise (PSNR = 32.0 dB) (B). (C) Noisy color image of an ostrich with 10% additive Gaussian noise. (D) Noise reduced output of the NV algorithm (PSNR = 32.7 dB). Pictures printed with permission from IEEE (see Liu et al. [4]).

patchy regions of homogeneous color, when the same noisy image was used as input. Furthermore, GSM wavelet denoising of this image resulted in image blur.

4. Discussion

The first two image-processing steps of this novel bionic NV algorithm were inspired by the transducer gain of photoreceptor cells of nocturnal insects and the spatial integration of image information in lamina neurons of *M. genalis* [2]. The combination of them allowed the strong enhancement of the contrast of dim images and effectively removed noise that would have been amplified after performing histogram stretching (see **Figures 2** and **3**). A drawback of many denoising algorithms is obvious by the loss of fine image details and sharp object contours, which leads to image blur and staircase effects. In contrast, spatial domain-filtering by means of 'adaptive spatial averaging' showed an unexpected denoising performance, which removes noise without affecting fine image details and object contours. The denoising performance of the NV algorithm is comparable to the performance of well-known denoising techniques such as wavelet, bilateral, TV, ATVM, BM3D filtering as long as the noise level of the input image remains below a critical value. If noise dominates the input image, the output image may contain small artifacts in the form of grain noise in homogeneous image regions (see **Figure 2C**).

Automatic parameter estimation as described here was sufficient to improve the quality of 12 pictures having different average brightness and noise levels. The method used here for noise estimation is similar to that described by Förstner [24] who estimated the noise level from the gradient of smooth or fine-textured regions, whereby these authors estimated signal-dependent noise level for each intensity interval. Similarly, Stefano et al. [25] described three methods to estimate the noise levels of natural images, and a noise estimation method based on mean absolute deviation was described by Donoho [26].

It will be possible to modify the bionic NV algorithm to improve the contrast of image sequences in dim films such as those generated by surveillance cameras operating at low light levels or digital night vision goggles. To reduce the noise of input images, it may be helpful to insert an additional processing step that performs a temporal summation of grey values by averaging across subsequent frames. A bionic method that operates in the spatial and temporal domains was described by Warrant et al. [5]. The night vision method described there is based on a smoothing kernel that is constructed for each pixel in three dimensions (two in space and one in time). In contrast to this complex algorithm, the NV algorithm described in this study is computationally less demanding and even runs on FPGA hardware due to the pixel-wise operations employed.

FPGA hardware can process images almost in real time due to its parallel architecture [19, 20]. To compute histogram statistics and equalization in parallel on a FPGA chip [27], non-conventional schemes for real-time histogram equalization have been discussed and implemented by Alsuwailem and Alshebeili [28]. Furthermore, several studies have investigated the implementation of the brightness control, contrast stretching, and histogram equalization algorithm on FPGA [29, 30], which has become a competitive alternative for high-performance digital signal processing (DSP) applications. Bittibssi et al. [31] addressed the hardware implementation of five image-enhancement algorithms in the spatial domain using FPGA: median filter, contrast stretching, histogram equalization, negative image transformation, and power law transformation. Recently, this NV algorithm was successfully implemented on a Trenz Electronic FPGA hardware platform for the purpose of denoising mammography images (prototype of the Mammobee project that was funded by the AWS Austria).

5. Conclusions

This bee-eyes inspired NV algorithm is based on rather simple calculations and operates in the spatial domain to suppress sensor noise in static images. It is applicable to a great variety of dim images differing in their brightness and degree of noise. Such a simple algorithm is suitable for FPGA technology, which allows image processing steps to be executed in parallel at the level of pixels. Since all parameters can be derived from the statistics of the input images, its use in the form of digital night vision goggles, medial applications and real-time fluorescence imaging systems is possible. In the future, this method will be adapted for night vision cameras that almost operate in the dark. For this task, subsequent video frames will be averaged to improve visual information. Another project is dedicated to the enhancement of dim X-ray images. This approach is based on the assumption that X-ray exposure during breast cancer screenings can be reduced by the amplification and denoising of slightly underexposed diagnostic raw data images. This is a challenging task because the resolution of such images as well as the range of grey values (16 Bit) is high and the procedure used for the noise estimation as well as the calculation of the threshold for "adaptive spatial averaging" needs to be modified.

Acknowledgements

The research was funded by the Austrian Science Fund P25709-B25.

Author details

Manfred Hartbauer
Institute of Biology, University Graz, Universitätsplatz 2, Graz, Austria

*Address all correspondence to: manfred.hartbauer@uni-graz.at

References

[1] Greiner B, Ribi WA, Warrant EJ. A neural network to improve dim-light vision? Dendritic fields of first-order interneurons in the nocturnal bee *Megalopta genalis*. Cell and Tissue Research. 2005;**322**:313-320

[2] Warrant E. Seeing in the dark: Vision and visual behaviour in nocturnal bees and wasps. The Journal of Experimental Biology. 2008;**211**:1737-1746

[3] Stöckl AL, O'Carroll DC, Warrant EJ. Neural summation in the hawkmoth visual system extends the limits of vision in dim light. Current Biology. 2016;**26**:821-826

[4] Liu C, Szeliski R, Bing Kang S, Zitnick CL, Freeman WT. Automatic estimation and removal of noise from a single image. IEEE Transactions on Pattern Analysis and Machine Intelligence. 2008;**30**:299-314

[5] Warrant E, Oskarsson M, Malm H. The remarkable visual abilities of nocturnal insects: Neural principles and bioinspired night-vision algorithms. Proceedings of the IEEE. 2014;**102**:1411-1426

[6] Buades A, Coll B, Morel J. A review of image denoising algorithms, with a new one. Multiscale Modeling and Simulation. 2005;**4**:490-530

[7] Nirmala SO, Dongale TD, Kamat RK. Review on image enhancement techniques: FPGA implementation perspective. International Journal of Electronics, Computer and Communications Technologies. 2012;**2**

[8] Rudin LI, Osher S, Fatemi E. Nonlinear total variation based noise removal algorithms. Physica D: Nonlinear Phenomena. 1992;**60**:259-268

[9] Lee YJ, Lee S, Yoon J. A framework for moving least squares method with total variation minimizing regularization. Journal of Mathematical Imaging and Vision. 2014;**48**:566-582

[10] Yoon SM, Lee YJ, Yoon G-J, Yoon J. Adaptive total variation minimization-based image enhancement from flash and no-flash pairs. Scientific World Journal. 2014;**4**:e319506

[11] Tomasi C, Manduchi R. Bilateral filtering for grey and color images. In: Proceedings of the IEEE International Conference on Computer Vision; 1998. pp. 839-846

[12] Zhang M, Gunturk BK. Multiresolution bilaterals filtering for image denoising. IEEE Transactions on Image Processing. 2008;**17**:2324-2333

[13] Maggioni M, Katkovnik V, Egiazarian K, Foi A. Nonlocal transform-domain filter for volumetric data denoising and reconstruction. IEEE Transactions on Image Processing. 2013;**22**:119-133

[14] Dabov K, Foi A, Katkovnik V, Egiazarian K. Image denoising with block-matching and 3D filtering. Proceedings of SPIE - Journal of Electronic Imaging. 2006;**6064**

[15] Crouse MS, Nowak RD, Baraniuk RG. Wavelet-based statistical signal processing using hidden Markov models. IEEE Transactions on Signal Processing. 1998;**46**:886-902

[16] Fan G, Xia X-G. Image denoising using a local contextual hidden Markov model in the wavelet domain. IEEE Signal Processing Letters. 2001;**8**:125-128

[17] Portilla J. Full blind denoising through noise covariance estimation using Gaussian scale mixtures in the wavelet domain. Proceedings of IEEE International Conference on Image Processing. 2004;**5**:1217-1220

[18] Bader DA, Jájá J, Harwood D, Davis LS. Parallel algorithms for image enhancement and segmentation by region growing, with an experimental study. The Journal of Supercomputing. 1996;**10**:141-168

[19] Zhang N, Chen Y, Wang J. Image parallel processing based on GPU. In: 2nd International Conference on Advanced Computer Control (ICACC); 2010. pp. 367-370

[20] Russo F. Piecewise linear model-based image enhancement. EURASIP Journal on Advances in Signal Processing. 2004;**12**:1861-1869

[21] Rao DV, Patil S, Babu NA, Muthukumar V. Implementation and evaluation of image processing algorithms on reconfigurable architecture using c-based hardware description language. International Journal of Theoretical and Applied Computer Sciences. 2006;**1**:9-34

[22] Xilinx Inc. System Generator for Digital Signal Processing. 2012. Available from: http://www.xilinx.com/tools/dsp.htm

[23] Portilla J, Strela V, Wainwright MJ, Simoncelli EP. Image denoising using scale mixtures of Gaussians in the wavelet domain. IEEE Transactions on Image Processing. 2003;**12**:1338-1351

[24] Förstner W. Image preprocessing for feature extraction in digital intensity, color and range images. LNES. 2003;**95**:165-189

[25] Stefano A, White P, Collis W. Training methods for image noise level estimation on wavelet components. EURASIP Journal on Advances in Signal Processing. 2004;**16**:2400-2407

[26] Donoho D. De-noising by soft-thresholding. IEEE Transactions on Information Theory. 1995;**41**:613-627

[27] Salcic Z, Sivaswamy J. IMECO: A Reconfigurable FPGA-Based Image Enhancement Co-Processor Framework. IEEE TENCON-Speech and Image Technologies for Computing and Telecommunication, TENCON '97. Brisbane, Australia; 1997. pp. 231-234

[28] Alsuwailem AM, Alshebeili S. A new approach for real time histogram equalization using FPGA. In: Proceedings of the International Symposium on Intelligent Signal Processing and Communication Systems; 2005. pp. 397-400

[29] Sowmya S, Paily R. FPGA implementation of image enhancement algorithm. International Conference on Communication and Signal Processing (ICCSP); 2011. pp. 584-588

[30] Acharya A, Mehra R, Takher V. FPGA based non uniform illumination correction in image processing applications. International Journal of Computer Technology and Applications. 2011;**2**:349-358

[31] Bittibssi TM, Salama GI, Mehaseb YZ, Henawy AE. Image enhancement algorithms using FPGA. International Journal of Computer Science and Communication Networks. 2012;**2**:536-542

Chapter 4

Pre-Harvest and Post-Harvest Techniques for Plant Disease Detections

Maki K. Habib and Hashem Rizk

Abstract

As the agriculture industry is growing fast, many efforts are introduced to ensure a high quality of produce. Diseases and defects found in plants and crops affect greatly the agriculture industry. Hence, many techniques and technologies have been developed to help solve or reduce the impact of plant diseases. Imagining analysis tools and gas sensors are becoming more frequently integrated into smart systems for plant disease detection. Many disease detection systems incorporate imaging analysis tools and VOC (Volatile Organic Compound) profiling techniques to detect early symptoms of diseases and defects of plants, fruits, and vegetative produce. These disease detection techniques can be further categorized into two main groups: preharvest disease detection and postharvest disease detection techniques. This paper aims to introduce the available disease detection techniques and to compare them with the latest innovative smart systems that feature visible imaging, hyperspectral imaging, and VOC profiling. In addition, this paper considers the efforts to automate imaging techniques to help accelerate the disease detection process. Different approaches are analyzed and compared in terms of work environment, automation, implementation, and accuracy of disease identification along with the future evolution perspective in this field.

Keywords: preharvest, postharvest, disease detection, plants, fruits

1. Introduction

The agriculture industry is undoubtedly one of the most vital sectors contributing to the national income of many developing countries. Throughout the years, many agriculture components and processes have become automated to ensure faster production and to ensure products of the highest quality standards. Because of the increased demand in the agricultural industry, agricultural produce must be cultivated using an efficient process [1]. Diseases and defects found in plants and crops have a great impact on production in the agriculture industry and can lead to significant economic losses [2]. A loss of an estimated 33 billion dollars every year was the result of plant pathogens found in crops in the United States. Pathogenic species affect plants significantly, introducing diseases such as chestnut blight fungus and huanglongbing citrus greening disease [3]. Insect infestation along with bacterial, fungal, and viral infections are other main contributors to diseases found in plants [4]. Changes in climate and temperature are also factors that may contribute to the increase in diseases found in plants. Once a plant has been infected, symptoms develop on various segments of the plant, ultimately degrading the growth of the subsequent fruit or vegetable [5].

Apple production is a huge industry especially in China with over 17 million tons of produce every year. Apple infections do not only significantly reduce grade and yield, but can also affect the return bloom of the following season [6]. These infections have drastic impacts on countries that rely heavily on their agriculture sector as their main source of income. In order to overcome these losses and issues of plant diseases, farmers tend to use chemical pesticides as a remedy solution. This solution may be effective in eliminating plant diseases but has drastic drawbacks. As well as being costly, the increased use of pesticides creates dangerous toxic residue levels on agricultural produce [7]. Not only does the toxic residue affect the healthiness of agricultural produce, but also has a significant impact on the surrounding environment seeping through the soil and into groundwater. This leads to concerns raised by the public about the wholesomeness and healthiness of products when pesticides are commonly used in the produce they purchase [8]. Therefore, the use of pesticides must be controlled and used only when necessary. This controlled or monitored method of pesticide use is known as selective pesticide spraying.

Many techniques have been introduced in order to decrease losses found in defective plants. Manual techniques, such as hand inspection and naked eye observation are very common methods used by farmers. Plant diseases are detected and characterized by observation from experts, which can be very expensive and time-consuming [2]. Because these methods are very tedious it is prone to sorting errors and judgmental errors from different farmers [6]. Therefore, different disease detection systems were introduced to tackle many of the issues faced with labor-intensive techniques.

A disease detection system has the ability to not only detect early symptoms of defective plants but can also prevent the disease from spreading. Disease detection techniques can be categorized into two methods: direct and indirect methods [4] as shown in **Figure 1**. Direct detection techniques rely on the use of laboratory-based experiments. The most popular and commonly used experiments are enzyme-linked immunosorbent assay (ELISA) and polymerase chain reaction (PCR) [4]. Indirect methods rely on advanced techniques that are mainly focused on field-based approaches. Indirect methods emphasize the integration of sensors and imaging techniques on-site, to provide a rapid and accurate method for disease detection. Early detection of apparent diseases in plants is of utmost importance [1], as this will aid farmers to take appropriate precautions to help preserve the defective plant. Indirect methods are vast and can be used for disease detection in both preharvest plants and post-harvest produce. If early detection is possible, the percent of defective fruits can be significantly decreased, while maintaining high-quality production standards.

1.1 Direct methods

When a pathogen attacks a plant, the plant DNA is altered, and a specific type of protein molecules are introduced to the plant by the pathogen during infection [9]. Direct methods focus on molecular and serological techniques that test the biological structure of the plant to check for the pathogen DNA or the presence of pathogen produced protein molecules. Commonly known techniques are the Polymerase chain reaction (PCR) and the Enzyme-linked immunosorbent assay (ELISA). PCR-based disease detection involves genetic material (DNA) extraction of the microorganism causing the disease. The gel electrophoresis is then performed after the DNA has been purified and amplified. If a specific band is present in the gel

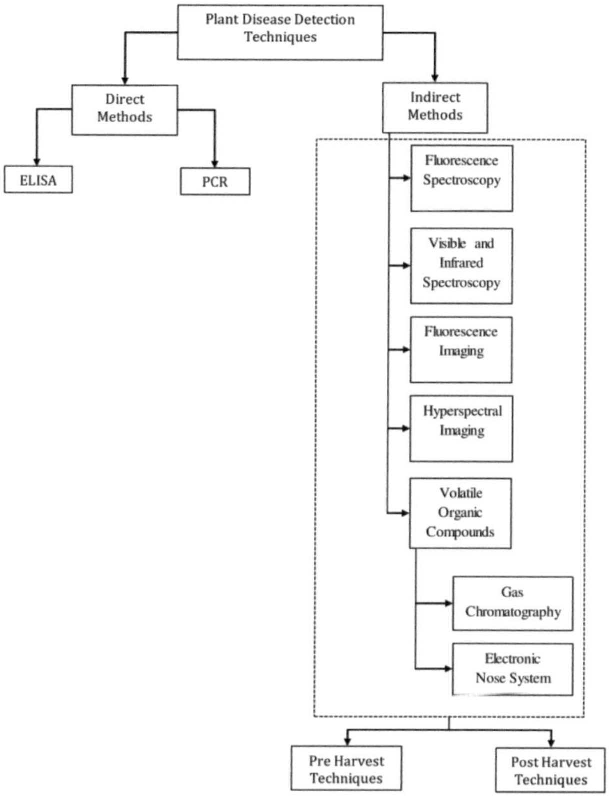

Figure 1.
Different methods of plant disease detection.

electrophoresis, then the existence of the plant disease organism is verified [4]. ELISA works by injecting the microbial protein of a specific plant disease into an animal, which in return produces antibodies against that specific disease. The extracted antibodies are used alongside fluorescence dye and enzymes for disease detection. If the plants were infected, then the sample would fluoresce, verifying the presence of a particular plant disease [4]. **Table 1** illustrates the difference between each technique and how they compare in disease detection. Because of these techniques, diagnostic kits have been designed to successfully detect diseases in crops such as rice and can also identify genetically modified organisms (GMOs) in shipments of conventional crops.

Although these techniques may be robust and very accurate in detecting plant diseases, the drawbacks of these methods are significantly vast. These techniques rely heavily on the use of expensive laboratory equipment and extensive experiments, which can be time-consuming and labor-intensive. Sample preparation

	ELISA	PCR
Diagnostic Kit	Protein-based	DNA-based
Cost	Simple laboratory equipment with no training required	Expensive costly equipment
Disease Detection	Root crops, fruits, and grains	Bananas, potatoes, and cotton

Table 1.
A comparison between ELISA and PCR techniques for plant disease detection.

consumes a considerable amount of time and effort to ensure reliable and accurate results. These techniques are also very expensive because of the use of consumable reagents that are specifically designed for each pathogen [4]. Therefore, new and more rapid disease detection systems are needed as a preliminary screening tool for processing large numbers of plant samples.

1.2 Indirect methods

New automated non-destructive methods have been studied to detect plant disease symptoms early and with high sensitivity to specific diseases. The most common techniques used are spectroscopic and imaging techniques for the detection of symptomatic and asymptomatic plant diseases [4]. The methods studied include fluorescence spectroscopy, visible and infrared spectroscopy, fluorescence imaging, hyperspectral imaging, and VOC profiling.

Fluorescence spectroscopy works by exciting an object with a beam of light and measuring the fluorescence released. Fluorescence spectroscopy systems can be used in field-based settings where leaves are still attached, and in laboratory settings where sample leaves are detached. Studies have shown that this method is accurate in discriminating defective plants from non-defective plants [10]. However, studies have also shown that this method is inefficient in providing enough information, such as water stress levels [11]. Visible and infrared spectroscopy is a rapid and cost-effective technique that can be used as a plant disease detection system. Studies have proven that this method can be used to detect stress levels and detect plant diseases accurately, even before symptoms are visible [12]. Unlike fluorescence spectroscopy, where only a single spectrum is used, fluorescence imaging works by using images obtained from a camera. Wavelengths from an object are recorded on a camera after fluorescence excitation from a UV light. A study confirmed that this method could be used to detect the tobacco mosaic virus (TMV) in tobacco plants.

Non-infected tobacco plants were successfully differentiated from infected ones in a relatively short period of time [13]. Hyperspectral imaging is similar to multispectral imaging but uses a wider range of wavelengths for each pixel. An image is produced consisting of a set of pixel values at each wavelength of the spectra. This method is common in monitoring the quality of food products, such as fruits. A study developed a system for early detection of yellow rust disease in winter wheat using visible-NIR hyperspectral imaging. Using quadratic discriminant analysis as the classification method yielded an accuracy of 92–98% when classifying diseased plants [14]. VOCs are released by plants due to factors such as humidity, temperature, light, soil condition, and fertilization [15] Studies have shown that certain compounds are released when a plant is under stress or is infected by a particular disease [16].

These methods were proposed to be integrated with an autonomous agricultural vehicle to provide real-time feedback of plant stresses and nutrient deficiencies. Not only have these methods been shown to provide successful detection of plant stress and nutrient deficiencies, but also have been useful in monitoring the quality of postharvest fruits and vegetables.

2. Preharvest disease detection techniques

Indirect methods can be used to detect diseases in both preharvest and postharvest fruits/vegetables. To have a better understanding of different disease detection techniques, indirect techniques can be further categorized into two main groups: preharvest and postharvest detection techniques. This section will include more in-depth research on the use of imaging techniques and VOC detection techniques for disease detection in preharvest fruits/vegetables.

Patil [17] proposed a method in which leaf disease severity of preharvest sugarcane leaves can be measured using imaging techniques. Fungi diseases are very common in sugarcane leaves and inhibit their growth immensely. These diseases leave visual spots on leaves, which in turn prevent the vital process of photosynthesis. Photosynthesis is a fundamental process essential for growth and prosperity. Rather than using pesticides, which is not only costly but also increases toxic residue levels, an early disease detection system can be implemented. Because fungi-caused diseases in sugarcane leaves appear as spots it is applicable to use imaging techniques to detect the severity of the disease [17]. Disease severity is expressed as the ratio between the affected area and the leaf area. If the lesion area to leaf area ratio is high, then the leaf is said to have a high disease severity according to **Table 2**.

For this study, 90 infected sugarcane leaves were used as image samples taken by a 12 Megapixel digital camera. The images were taken in a controlled environment with a white background and light sources to eliminate any reflection and to provide enhanced view and brightness [17]. For improved results, the input leaf image in **Figure 2** is first transformed from RGB color space to HIS color space. The image is then converted to grayscale as shown in **Figure 3** which is then segmented into two regions: the diseased region and the healthy region shown in **Figure 4**.

In order to segment the image, a triangle thresholding method was used. After the image has been segmented the leaf area and infected area ratios are calculated by measuring the number of pixels in the white region and the black region, respectively. This experiment showed to be very successful with an average accuracy of 98.60%.

To further improve the efficiency and accuracy of leaf disease grading systems, Sannakki [7] proposed the use of machine vision and fuzzy logic for disease detection in pomegranate leaves. Similar to the previous research, a table-like

Category	Severity
0	Apparently Infected
1	0–25% Leaf Area Infected
2	26–50% Leaf Area Infected
3	51–75% Leaf Area Infected
4	>75% Leaf Area Infected

Table 2.
Disease severity scale developed by Horsfall and Heuberger [17].

Figure 2.
RGB color space of brown spotted diseased sugarcane leaf [17].

Figure 3.
Grayscale image of brown spotted diseased sugarcane leaf [17].

Figure 4.
Infected region detection after triangle thresholding [17].

percent-infection calculation method is used to illustrate the severity of the disease. A 10 megapixel Nikon Coolpix L20 digital camera was used to take images of disease-infected pomegranate leaves at an equal distance of 16 cm. Images were taken from several pomegranate farms with no specific test site indication. A few image restoration techniques are implemented before the image is segmented into two specific regions: disease region and healthy region. Firstly, to reduce the computational complexity of the system, images are first resized to a fixed resolution. Then by the use of a Gaussian filter any noise or outer interference in the image is removed or diluted. K-means clustering (K = 10) technique is used here as the segmentation method, and the cluster that contains the diseased spots are saved to calculate the area of the diseased region. After calculating the disease area (AD) and total leaf area (AT), it is now possible to calculate the percent of infection in regard to the total area. A fuzzy logic system is implemented in order to characterize which disease grade the disease belongs to depending on the percent infection. It is difficult to assess which disease grade the disease belongs to because of the ambiguity and uncertainty of the table, but a fuzzy logic system can be effective in this case. The segmented regions that are used to calculate the percent-infection index are shown in **Figures 5** and **6**. The fuzzy logic system proved to be very effective in accurately grading diseases into their appropriate categories.

Because it can take up to one week to diagnose plant diseases using traditional chemical analysis, Xu [18] proposed an early plant disease detection technique. During the period of anthesis plants often appear to be nutrient deficient and it is vital to detect these deficiencies early to ensure the quality and quantity of plants. Nutrient deficient plants usually leave quite visible symptoms on their leaves, which can be

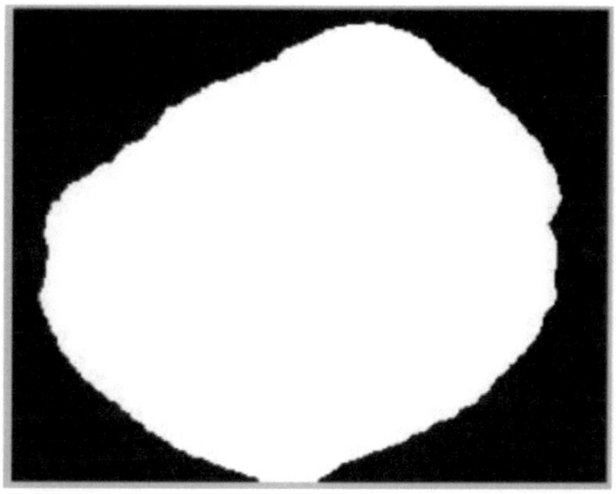

Figure 5.
Total leaf area (AT) of pomegranate leaves shown as a white spot [7].

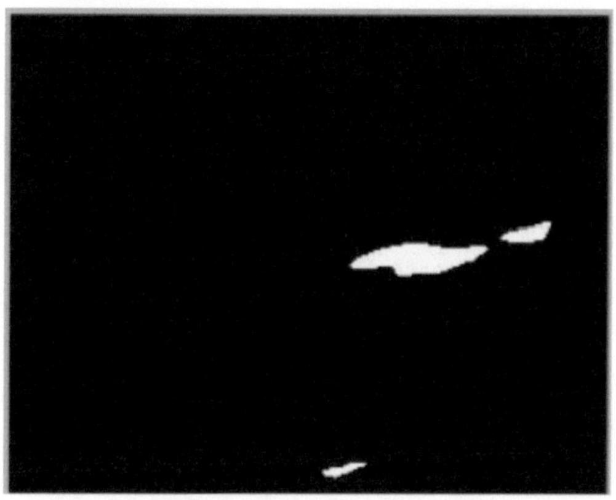

Figure 6.
Disease area (AD) of a pomegranate leaf shown as white spots [7].

used to diagnose the disease. By extracting features from leaves such as color and texture, plant nutrient deficiency can be diagnosed at an early stage [18].

Nitrogen and potassium deficient tomato plants were used under a controlled chamber environment to extract color and texture features of leaves. A sampling box shown in **Figure** 7 was used to take colored images of tomato leaves with a digital camera. Normal, prophase, and anaphase of nitrogen and potassium deficient tomato leaves were used in the proposed system as shown in **Figure** 7.

After images are taken in the chamber they are processed for color and texture feature extraction. Because nitrogen-deficient leaves turn yellow with the development of a disease, the number of yellow pixels in the leaf image can reflect the feature of nitrogen deficiency. Therefore, Xu [19] proposed the use of percent histograms to extract color features. The extracted feature set was classified using a fuzzy K nearest neighbor classifier and proved to be quite accurate in identifying normal, nitrogen deficient, and potassium deficient leaves.

The proposed system was tested using sample collection of tomato leaves and showed to be effective with an accuracy of identifying normal leaves, nitrogen-deficient leaves, and potassium deficient leaves to be 92.5%, 85%, and 82.5% respectively. The diagnostic system can identify potassium deficiency in leaves 10 days earlier than by experts' observation. This gives a significant amount of time for measures to be taken to ensure high-quality production.

Figure 7.
Examples of images used in the diagnostics system [18].

Figure 8.
Image of cotton crops showing the visual symptoms of damages caused by (a) southern green stink but; (b) bacterial angular; (c) Ascochyta blight [20].

In order to help crop producers and farmers in remote areas to identify early symptoms of plant disease, Camargo [20] used image processing and pattern classification to implement a machine vision system. The system would detect cotton crop damage caused by 3 specific diseases: green stink bug, Bacteria angular, and the Ascochyta blight virus as shown in **Figure 8**. A set of 117 images were used in the study from different sections of the plant such as the leaf, fruit, and stem [20]. Multiple features that characterize the shape and appearance of the image were extracted using both the image's RGB and HSV color space. For texture extraction, a co-occurrence matrix was used to identity gray levels between a specific position in the image and its neighboring pixel.

A box-counting algorithm was used to estimate the dimensions of the image for fractal dimension feature extraction. Gliding Box Algorithm was used to calculate the lacunarity feature to measure texture associated with patterns. A Support Vector Machine Classifier was used to identify the best classification model for the different feature sets. Each feature was used individually and grouped to identify the difference in classification accuracy with respect to feature type [20].

Results showed that grouping features of similar type resulted in higher classification accuracies when compared to using individual features. Results also showed that grouped texture features achieved higher classification rates (83%) when compared to grouped shape features which achieved a classification rate of 55%. When using the total feature set the highest classification rate was achieved of 90%. To further enhance the classification accuracy, one feature was withdrawn at a time until the best feature set was discovered. Using this method, a deducted feature set resulted in a classification accuracy rate of 93.1%.

In order to decrease the cost of uniform and periodical pesticide spraying, Oberti [8] suggested a selective spraying solution using an agriculture robot. Because primary infections start from localized discrete foci, a system can be implemented to detect these infected foci points and targeted treatment can be established. Eliminating the initial infection point does not only inhibit the spread rate of the infection to other patches but also significantly decreases the use of pesticides. In this case study, a multispectral camera is used to inspect an entire grapevine canopy for automatic detection of powdery mildew and selective spraying is achieved using a six-degree robotic arm illustrated in **Figure 9**.

The red, green, and NIR channels were the primary channels used for disease detection. An RGB camera was also added to the camera rig for visual documentation of the scene. The cameras were positioned at a constant height of 1.4 m while maintaining a distance of 1 m from the canopy wall as shown in **Figure 10**. Halogen light panels were used to provide diffused illumination of the imaged area for enhanced and more accurate results for indoor conditions. The disease detection algorithm used in this system was to capture very sharp changes in the reflectance in green and red channels, as this will give a clear indication of the breakdown in chlorophyll content in infected leaves. Specifically, two indexes were calculated and used to measure chlorophyll absorption illustrated in Eqs. (1) and (2):

$$I1 = (\text{Red Green}) = \text{NIR}^2 \tag{1}$$

$$I2 = \text{Red} = (\text{Red} + \text{Green} + \text{NIR}) \tag{2}$$

Since healthy regions have high chlorophyll absorption, it is expected that I1 and I2 will yield higher values for diseased areas and return lower values for healthier regions. To test the proposed system, grapevine plants were used to recreate a vineyard canopy wall in a greenhouse as shown in **Figure 11**.

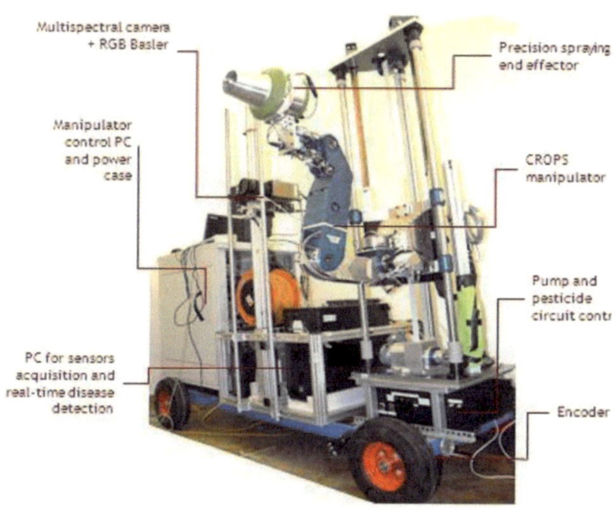

Figure 9.
Robotic vehicle for disease detection and selective pesticide spraying [8].

Figure 10.
Camera setup for the agricultural robot [8].

Figure 11.
Healthy grapevine plants aligned together with infected plants positioned randomly circled in red [8].

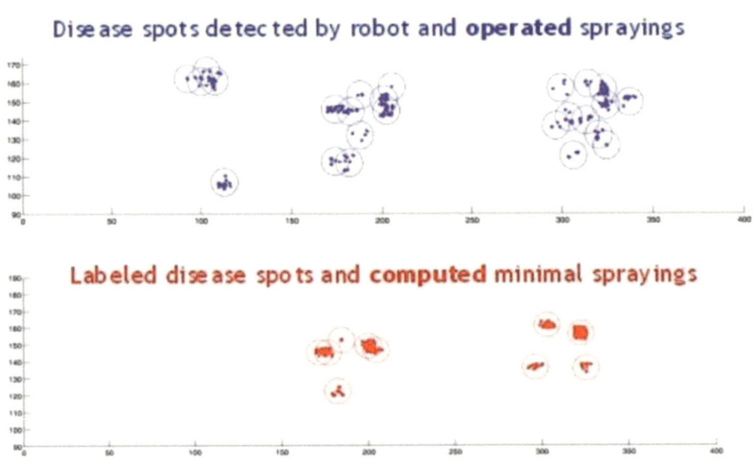

Figure 12.
Disease spots detected by robot and operated sprayings (blue) against labeled disease spots and computed minimal sprayings (red) [8].

Figure 12 illustrates the results gathered after the first experimental run. The blue graph shows the disease spots detected by the robot (blue dots) and the robot sprayings (blue circles) operated by the robot. To analyze the results, the blue spots are compared against the actual disease spots (red dots) specified by a plant pathologist and computed minimal sprayings (red circles). In the first experimental run, the robot detected all the diseased areas and successfully covered all the disease foci with pesticide spraying. The selective pesticide spraying achieved here obtained a reduction in pesticide use up to 84% when compared to homogenous spraying of the canopy. However, at different instances the robot did detect disease spots in areas where plants were healthy, this is apparent around the 100th mark. This false detection can be the result of illumination changes and shadow effects, which distort the input image and hence the values of I1 and I2. Also, the increased operated sprayings around detected diseased areas by the robot is apparent when compared to minimal sprayings. Oberti [8] claims that the surroundings of detected disease areas can be treated anyway by including a conservative safe-border area. This may help to raise the level of acceptance in real-world cases, despite the reduction in potential pesticide savings. Overall, the proposed system proved to detect disease foci with an accuracy of 85%, while achieving a reduction of pesticide use close to 90%.

Pobkrut [21] used a semi-autonomous mobile e-nose robot to examine the fertility of the soil by using metal oxide gas sensors to detect organic volatile compounds found in soil. A six-wheel robot was designed with an integrated array of gas sensors shown in **Table 3**. This e-nose system was used because of its low cost and high sensitivity to certain target gases [21].

Multiple fans were installed on the robot to ensure airflow towards the gas sensor enclosed chamber as shown in **Figure 13**. The data collected from the gas

Sensor	Target gases
TGS 825	Hydrogen sulfide
MQ2	Combustible gases
MQ5	LPG, natural gases
MQ135	NH3, NOx, Benzene, CO_2 gas
TGS 2600	Air contaminants (ethanol, isobutane, hydrogen)
TGS 2602	VOCs and odorous gases

Table 3.
TGS and MQ gas sensors and their target gases [21].

Figure 13.
A six-wheel robot with E-nose and navigation system and an e-nose chamber [21].

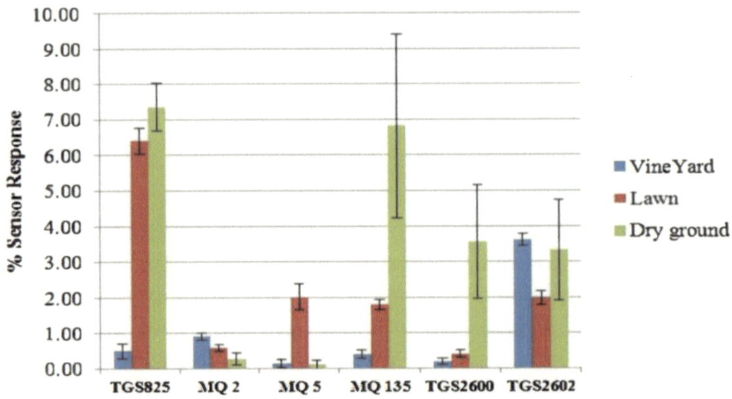

Figure 14.
Percent sensor responses of six elements used in the e-nose robot system to soil volatiles at different places [21].

sensors were recorded in real-time and sent to a computer for analyzing and visualizing via Zigbee wireless network. Ultrasonic sensors and accelerometers were also implemented in the robot to ensure obstacle avoidance and smooth navigation. The robot was operated under real conditions in four different locations, floor room, lawn, dry ground, and vineyard row.

From **Figure 14**, it can be observed that the results collected from the dry ground location yielded a high percentage in sensor response to most of the gas sensors. This can be because little to none of the volatile gases are absorbed from the surroundings because of the lack of weeds and grass [21]. Most gas sensors yielded quite low responses in the vineyard location except for TGS2602, because of its high sensitivity to odorous gases such as ammonia and hydrogen sulfide. TGS2602 also has a high sensitivity to VOCs such as toluene [21]. The six different sensors show promising results in indicating different volatile gases in dry ground. Pobkrut [21] argue that if enough common odor data from various places are collected and put into a database, this database can later be used to determine irregular events.

3. Postharvest disease detection techniques

After reviewing different approaches to plant disease detection in preharvest harvest produce it is necessary to discuss postharvest detection techniques.

The method proposed by Jhuria [1] uses image techniques and artificial neural networks (ANNs) to classify different diseases found in grapes and apples. Image processing was used to extract specific features such as fruit color, texture, and morphology. An important factor that may aid or diminish the effectiveness of image processing is the selection of the color space. Jhuria, Kumar, Borse [1] proposed that the HIS color space is more suitable than RGB as it is less affected by changes in light. A neural network was used to characterize these features into a disease category such as apple scab or apple rot. The neural network was first trained with a data set of various apple diseases. **Figure 15** illustrates the different steps taken during the training and testing of the neural network.

After being trained, the neural network was capable of characterizing an input apple image into its corresponding disease category. It was concluded that the selection of features plays a vital role in the effectiveness of the neural network. Because diseases are better defined by color and morphology, these features, unlike texture, proved to provide improved results [1].

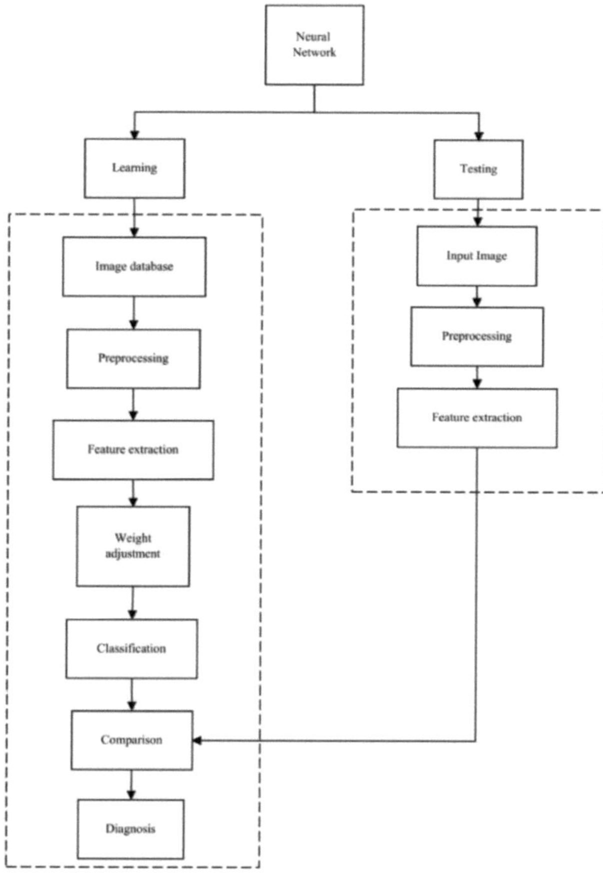

Figure 15.
Flowchart of the proposed system for training the neural network to determine diseases in apples and grapes [7].

Dubey [2] provided a similar technique with the use of a Multi-class SVM (Support Machine Vector) classifier and K-Means image segmentation to detect three diseases found in apples: apple blotch, apple rot, and apple scab. Because of the wide variety of skin colors found in different apples, it was proposed that using size and color as features could be challenging in the detection of defects in apples. Therefore, a K-means based image segmentation approach was used to extract disease features.

In order to ensure shorter processing times for image segmentation, L*a*b* color space was used [2]. K-means clustering was used to segment the apple image into 4 different clusters by categorizing similar pixel densities into their corresponding cluster. Dubey [2] proposed that for improved segmentation results, using 3 to 4 clusters was sufficient. **Figure 16** demonstrates the results of the K-means clustering for apples with different infections.

It is clear that the K-means clustering technique is an effective tool for image segmentation. Four different techniques were used for feature extraction, Global Color Histogram (GCH), Color Coherence Vector (CCV), Local Binary Pattern (LBP), and Complete Local Binary Pattern (CLBP). These color and texture features were used to validate the accuracy and efficiency of the proposed system [6]. GCH represents the probability of a pixel being a certain distinct color in the form of a histogram, whereas the CCV distinguishes coherent and incoherent pixels into two separate histograms. Coherent pixels are defined as pixels that belong to a large region with homogenous color, and any other pixel is defined as an incoherent

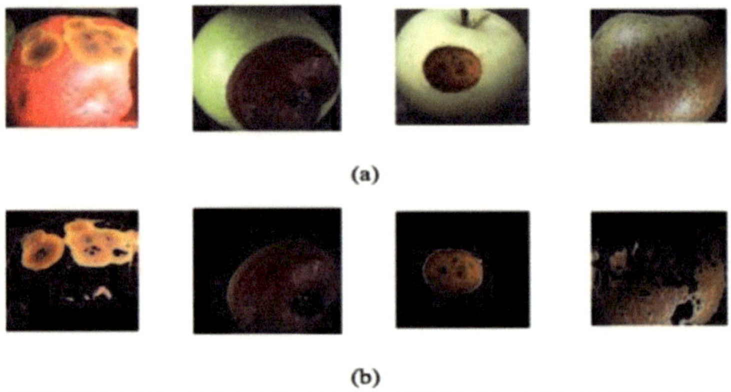

Figure 16.
Image results before (a) and after (b) K-means clustering segmentation [2].

pixel. LBP considers the difference of each pixel with respect to its neighbors. CLBP on the other hand, considers signs, magnitude, and original center gray level value of local differences. After the extraction of features, a Multi-class Support Vector Machine (MSVM) was used for the training and classification as shown in **Figure 17**.

Support Vector Machines have significant advantages over ANNs as they are less prone to overfitting and require less computational power. However, since ANNs use a heuristic method, it is easier to develop than an SVM, which involves more theory. A data set of 431 apple images was created with wide variations in apple type and color to ensure a more realistic test [2]. The data set is to be categorized into Apple Botch, Apple Rot, Apple Scab, or Normal Apple categories.

Results proved to coincide with Jhuria's [1] proposal that the use of HSV color space outperforms RGB color space [2]. It is clear from **Figure 18** that using the HSV color space in every feature extraction technique yields more accurate results. Also, **Figure 18** shows that the most accurate extraction techniques are the CLBP

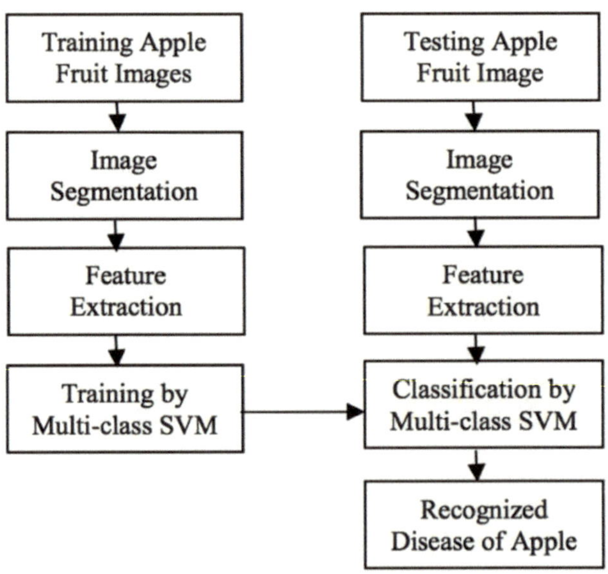

Figure 17.
Flowchart of the proposed MSVM system [2].

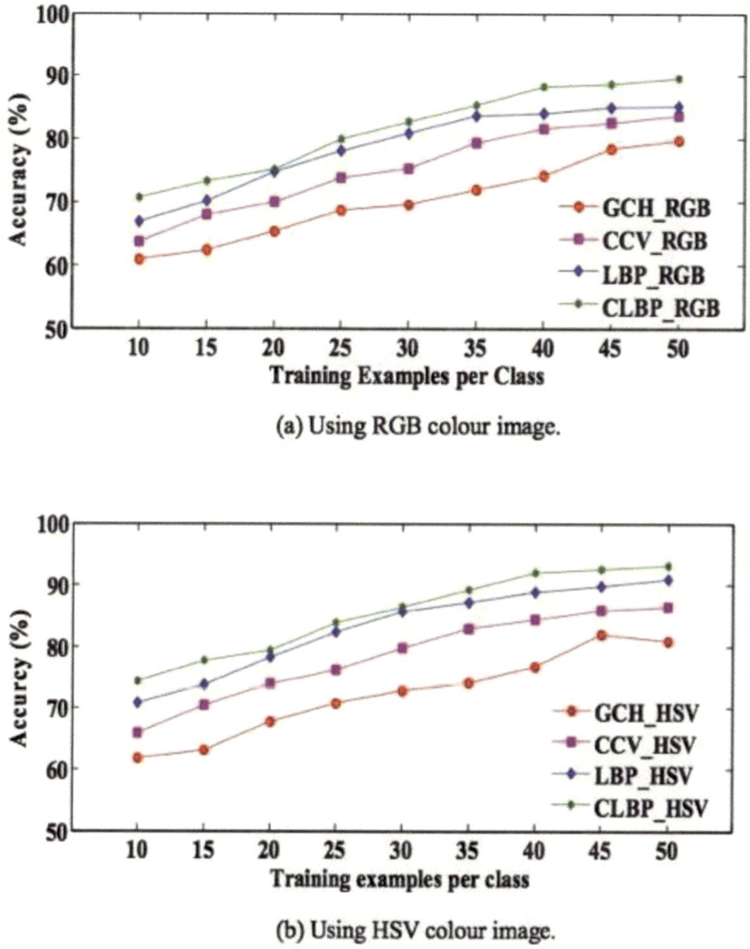

Figure 18.
Accuracy difference of using RGB color space compared to HSV [2].

followed by the LBP. Unlike GCH and CCV, both CLBP and LBP use information from neighboring pixels. Because they use local differences, they are more efficient in pattern matching and are less computationally extensive. It can be concluded that it is more effective to use either LBP or CLBP as feature extraction techniques to yield more accurate results. Furthermore, results indicated that the MSVM classifier detection of normal apples and apple scab was significantly easier than the detection of apple blotch and apple rot. **Figure 19** illustrates this observation, with very high accuracy results for the detection of normal apples and apple scab by using the LBP technique. However, the accuracy rates for the detection of apple blotch and apple rot are significantly lower.

To increase the speed of the sorting process, Unay [22] suggested a computer vision-based system to automatically grade apples. A monochrome digital camera with multiple band-pass filters was used to capture one-view images of 'Jonagold' apples taken in a controlled illuminated environment. The data set consisted of 280 healthy apples and 246 apples included several skin defects such as bruises and rot. The four bandpass filters used for image acquisition are centered at 450 nm (Blue), 500 nm (Green), 750 nm (Red), 800 nm (Infrared) as shown in **Figure 20**. Images of apples were taken in a uniform and low-intensity background to ensure a controlled environment. Therefore, background segmentation can be easily achieved

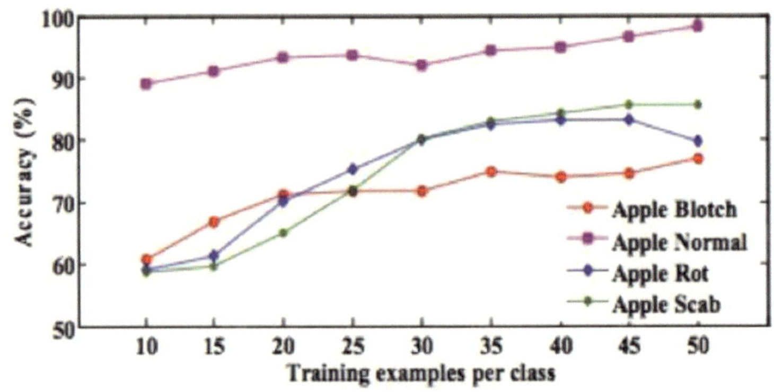

(a) LBP in RGB color space.

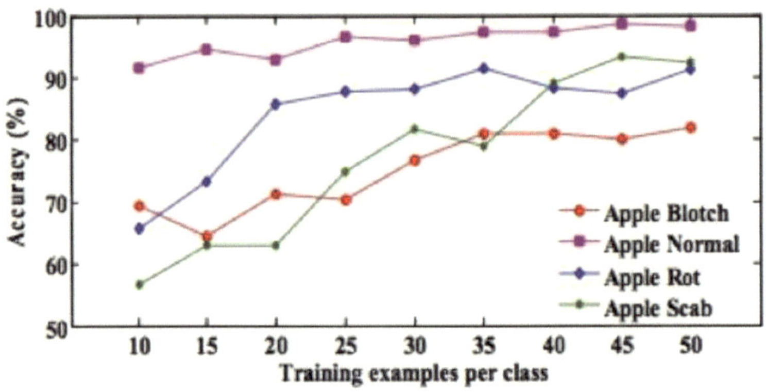

(b) LBP in HSV color space.

Figure 19.
Accuracy of detecting different apple disease categories in RGB and HSV color space [2].

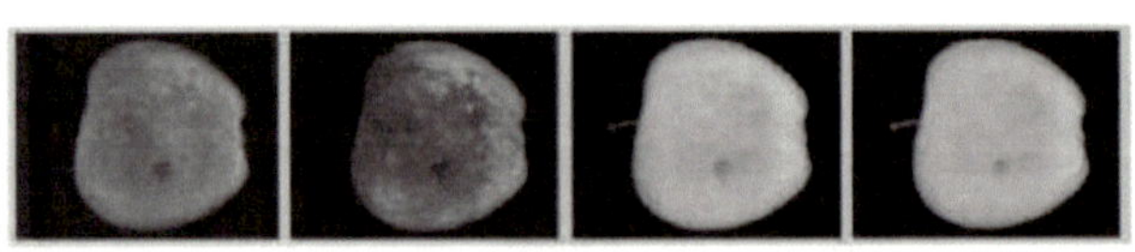

Figure 20.
Filter images of apple. Left to right: Blue, green, red, infra-red filters [22].

using a thresholding technique. Multiple global thresholding techniques such as Otsu, Entropy, and Isodata were tested for defect segmentation. However, because of the similar appearance of the stem-end/calyx area and the apple defect, a segmentation technique is required to distinguish them from one another. Stem-end and calyx are natural parts of the apple and usually appear as dark blobs which can be often mistaken as defects. Statistical, textural, and shape features are extracted and introduced to a support vector machine to distinguish the calyx from the defect. The result of this segmentation can be seen in **Figure 21**.

After calyx removal, a total of 13 features were extracted and introduced to multiple fruit classification techniques to test the accuracy of different classifiers. The 5 classifiers used were Linear Discriminant Classifier (LDC), Nearest Neighbor (k-NN), Fuzzy Nearest Neighbor (fuzzy k-NN), Adaptive Boosting (AdaBoost), and Support Vector Machine (SVM).

Figure 22 illustrates the different thresholding techniques with all 4 filters. It is quite apparent that unlike the blue and green filters, the red and infrared images

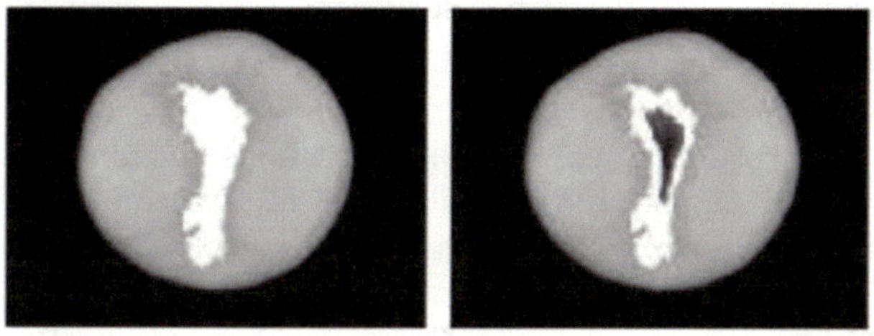

Figure 21.
Example of stem-end/calyx removal. Before the removal on the left and stem-end/calyx removed on the right. Defected areas displayed in white in both images [22].

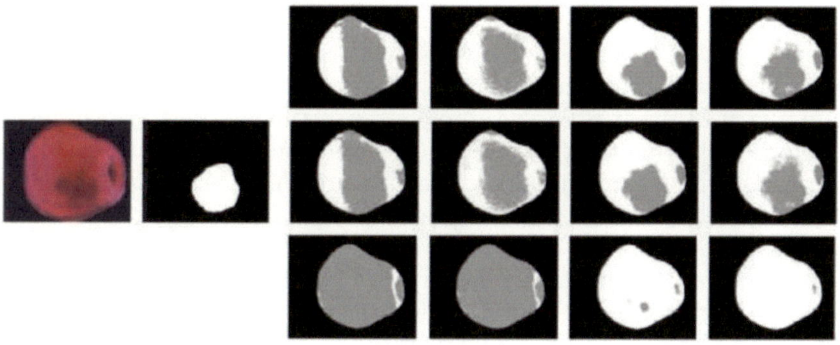

Figure 22.
Segmentation results of thresholding methods on a bruised apple. Original RGB image and the manual segmentation of the fruit are on the left. Subsequent synthetic images show defected regions in gray and healthy ones in white. Each row belongs to a thresholding method (top-to-bottom: Otsu, isodata, entropy) and each column shows a band (left-to-right: Bl, GR, RE, IR) [22].

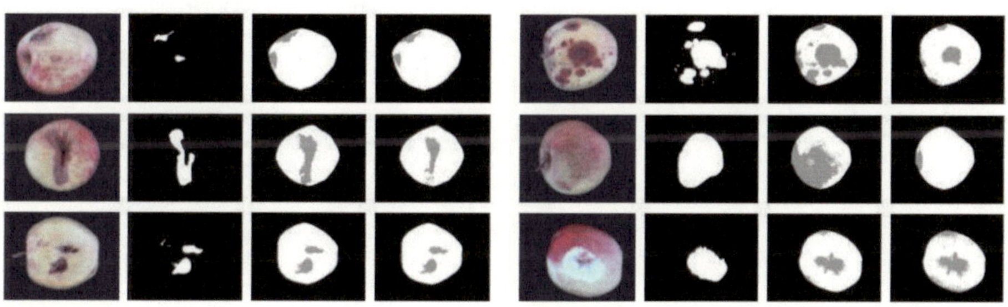

Figure 23.
Results of segmentation by isodata thresholding on RE and IR filter images. Fruits displayed are defected by scald (top-left), rot (top-right), frost damage (mid-left), bruise (mid-right), hail damage perfusion (bottom-left) and flesh damage (bottom-right). For each fruit, its original RGB image, its manual segmentation, and its segmentation results are displayed in a row. Defected areas are displayed in white in manual segmentation image, whereas segmentations show defected regions in gray color and healthy ones in white [22].

provide a more accurate representation of the defect segmentation. Blue and green filter images result in false segmentation because of the low contrast between healthy and defective skin in the wavelength range of 410-510 nm. **Figure 22** also shows that isodata thresholding accurately segments the defective area when compared to Otsu and entropy thresholding.

To further validate the results more apple images were segmented using the Isodata thresholding technique on Red and Infrared filters. **Figure 23** shows that results of Red filter images give better segmentation results when compared to Infrared filter images. After calyx removal, defect segmentation, and feature extraction, apples are graded by different classifiers as mentioned before. SVM proved to be the most accurate classifier (89.2%) in this case when using the Isodata method with the Red filter images. The LDC and k-NN classifiers performed lower with accuracies of 79% and 83% respectively.

Li [6] proposed a more automated approach using an on-line experimental system that can simultaneously inspect all four sides of an apple, and sort them accordingly. Two main issues with previous studies were tackled; the first is to acquire the whole surface of an apple at on-line speeds and to quickly identify the apple stem and calyx. A description of the system schematic is represented in **Figure 24**. The schematic displays how apples are fed into the machine vision system via conveyors and belts for image acquisition, and how they are sorted accordingly. The feeding conveyor is designed to ensure that the stem of the apple is faced upwards for maximum performance. The machine vision system consists of two cameras to provide multiple images of the apple, and a lighting chamber to control the light distribution [6]. By use of mirrors the top camera will cover three side views of the apple: top and two sides. The camera below will take an image of the bottom view of the apple. This setup has the distinctive advantage of inspecting all sides of the apple in one cycle. The setup is illustrated further in **Figure 25**.

After image acquisition, multiple methods were implemented for accurate defect detection. To not disrupt the segmentation process, image background removal algorithms were implemented to ensure that any backgrounds such as the mirror are removed. Segmentation is completed by using a reference fruit image and then subtracting it from the original fruit image. Then by the use of a simple thresholding method the defects could be easily extracted [6]. Because stem and calyx defects are very similar to each other the authors proposed the use of neural networks to distinguish the stem and calyx defects.

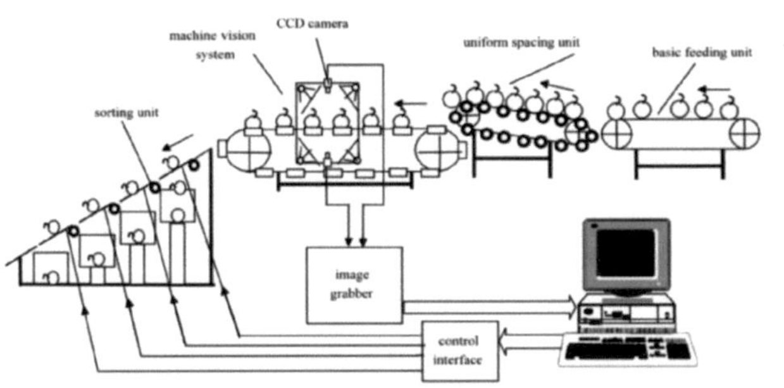

Figure 24.
Schematic representation of apple defects sorting system [6].

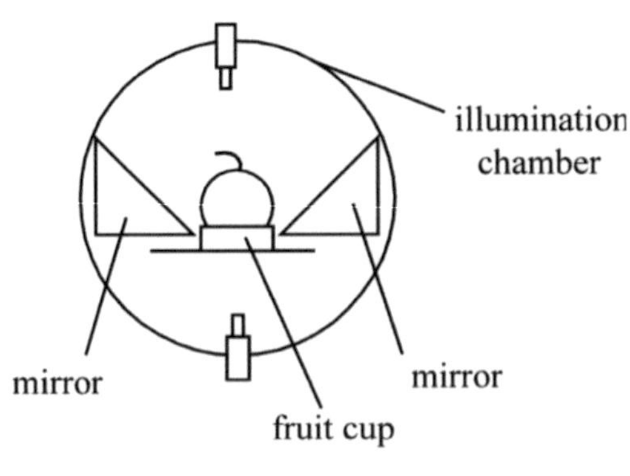

Figure 25.
Setup of the mirror vision system on the sorting system [6].

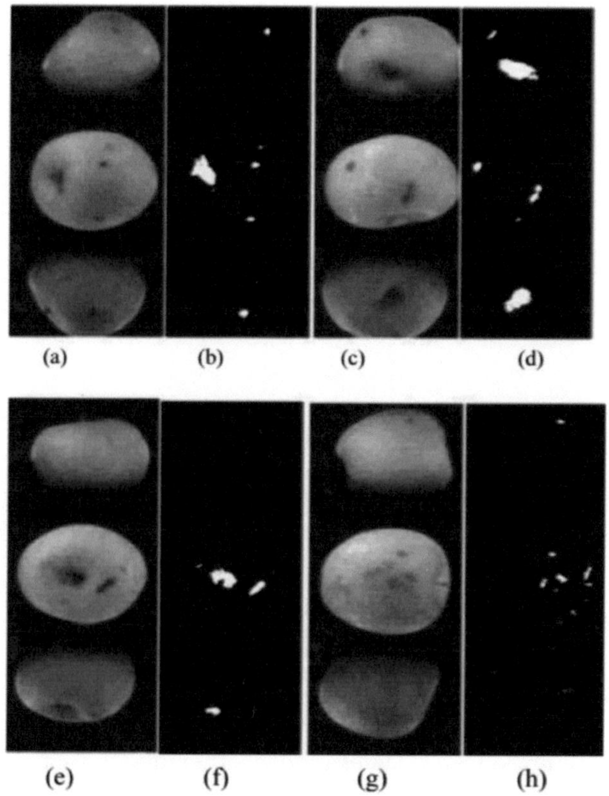

Figure 26.
Defects segmentation results. (a), (c), (e), and (g) original image; (b), (d), (f), and (h) segmented defects [6].

Forty samples of Fuji apples were used to test and validate the system. **Figure 26** illustrates how the input apple image is segmented and the defected regions as well as the stem and calyx defects are detected. The neural network classifier to be very effective in detecting stem-calyx recognition with accuracies over 93% [6]. Overall, this system proved to be successful in detecting defects on multiple sides of an apple simultaneously, while on a sorting line.

Xing [19] also implemented a non-destructive automated technique to detect bruises on apples using multiple waveband techniques. Hyperspectral imaging can provide enough information in several wavebands, but it is inappropriate in an on-line system due to its long acquisition and analysis time. Instead, Xing [19] resided on using multi-spectral imaging where only a specific range of wavebands was used to detect apple bruises.

A total of 128 'Golden Delicious' apples were purchased from two different locations and separated into non-bruised and bruised groups by visual inspection. Bruises are usually caused by human handling and vibration from transportation. Apples without visible bruises were impacted with a pendulum in the laboratory to achieve an average bruise size of 17 mm in diameter [19]. The hyperspectral imaging system consisted of a conveyor belt, light source, and camera for on-line bruise detection of apples as shown in **Figure 27**. The camera has high sensitivity from 400 to 1000 nm and is used alongside a spectrograph to detect the separate wavebands of light. The system also consisted of light sources and operated under a controlled environment to minimize interference from ambient light.

A simple thresholding method was used to segment the image background and low-intensity regions and further analyzed by principal component analysis (PCA).

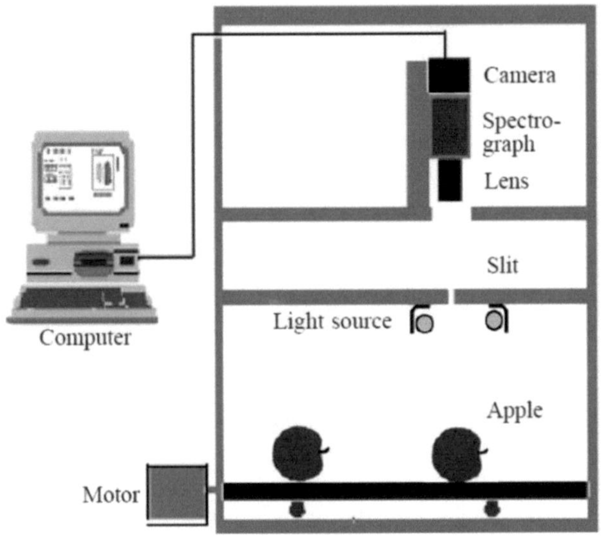

Figure 27.
Schematic of the used hyperspectral imaging system [19].

PCA is an effective tool in reducing data dimensionality and to enhance bruise features. Results showed that the wavebands centered at 558, 678, 728, and 892 nm were optimal in detecting bruises on 'Golden Delicious' apples. A simple classification technique was introduced to determine whether apples are bruised or intact. This classification technique resulted in an accuracy of 93.5% for detecting intact apples and about 86% for detecting bruised apples.

4. Conclusion

Both preharvest and postharvest imaging and VOC profiling techniques have proven to be very effective inaccurately classifying different types of plant diseases. Not only do these techniques give a good indication of overall plant health but also can accurately distinguish healthy produce from unhealthy produce. Preharvest techniques such as [18, 20] achieved a classification accuracy of plant diseases of 82.5% and 90% respectively. Postharvest techniques showed to be more promising as seen in [6, 22] with a classification accuracy of fruit defects of 89.2% and 93% respectively. Preharvest disease detection techniques can be classified as an early disease detection method as seen in [8, 18], in which immediate action can be taken to revive plants and crops. This aspect is a major advantage and cannot be achieved with postharvest techniques. Postharvest techniques such as [6, 19] have introduced automated systems in which defected postharvest produce can be distinguished and sorted automatically with an accuracy of 93% and 86% respectively. However, the integration of automation under real-time and field-based environments is still very limited.

As seen in [8], a robot rig is used under field-based conditions to detect diseases in grapevine canopy but is controlled manually and full automation is not achieved. Also, [21] implemented a semi-autonomous robot to test soil fertility under field-based environments using multiple gas sensors, but results were ineffective in disease detection.

The methods discussed differ greatly, from the use of simple digital cameras to the use of more advanced and sophisticated hyperspectral and multispectral imaging methods. Techniques such as [6, 19, 22] that use multispectral cameras and bandpass filters, show higher classification accuracy as seen in **Table 5** when compared to other techniques that use simple digital cameras for image capture. The use of multispectral cameras provides more information on the image that can be used to extract defected or diseased areas which may not be clear when using simple digital cameras that have low sensitivity in the higher wavebands [19]. indicated that higher wavebands were helpful in detecting bruises in apples, and this cannot be achieved when using a simple digital camera.

A. Key Achievements

- Plant nutrient deficiencies can be detected at a very early stage for fast and appropriate actions to be taken as shown in [18].
- Automated robotic selective pesticide spraying is achieved with adequate accuracy as shown in [8].
- The use of e-nose sensors and VOC profiling give farmers good indications on land and soil fertility as shown in [21].
- Automated sorting of fruits with very high accuracy is achieved as shown in [6, 19].

B. Challenges with Preharvest techniques

- Images of leaves are taken off the field and in controlled environments as shown in [7, 17, 18, 20].
- Automation techniques and the use of agriculture robots are not fully developed and achieve varying results as shown in [8, 21].
- Limited applications for processing large numbers of plants in real-time under field conditions.

C. Challenges with Postharvest techniques

- Tests are not field-based and conducted under controlled environments as shown with [1, 2, 6].
- Postharvest techniques are classified as late disease detection methods. It is difficult to cure the disease after the fruit has been fully developed.

5. Future work

After reviewing the research, multispectral and hyperspectral imaging techniques proved to be the most reliable indirect method. However, hyperspectral imaging is very costly, and there is still a limitation in the capability of designing systems for the detection of diseases in real-time under field conditions. Most preharvest and postharvest techniques are completed under controlled environments as seen in **Table 4** and automation techniques are developed mainly for postharvest produce for fruit sorting as seen in **Table 5**. Because the use of preharvest techniques provides an early analysis of disease severity, preharvest is more suitable to use over postharvest techniques.

Paper	Acquisition Method	Test Plant	Target Disease	Environment	Automation/ Manual	Coverage Area	Segmentation Method	Classifier	Accuracy
Patil [13]	12 Mega Pixel Digital Camera	Sugarcane leaves	Fungi diseases	Off-Site Controlled Background	Images Taken Manually	Single leaf	Triangle Thresholding	NA	98%
Li [9]	10 Mega pixel Digital Camera	Pomegranate leaves	General disease spots	Off-Site Controlled Background	Images Taken Manually	Single leaf	K-means clustering	Fuzzy Logic Classification	High
Xu [19]	Digital Camera with 0.4 million CCD pixels	Tomato plants	Nutrient deficiency	Closed Chamber	Images Taken Manually	Single leaf	Percent Histogram	Fuzzy k nearest neighbor classifier	82.50%
Camargo [4]	24 bit JPEG Image samples	Cotton Crop	Green stink bug, Bacteria angular, Ascochyta blight virus	Off-Site	Images Taken Manually	Single leaf	Co-occurrence matrix	Support Vector Machine Classifier	90%
Oberti [12]	Three-CCD Multispectral Camera 1912x1076	Grapevines	Powdery mildew	On-Site Controlled Lighting	Robot Rig Limited automation	Grapevine canopy	NDVI calculation	Local Gradient Method	85%
Pobkrut [15]	Multiple Gas Sensors (e-nose)	Soil fertility	VOCs	On-Site	Semi-autonomous six-wheel robot	Lawn	NA	NA	NA

Table 4.
Review of preharvest disease detection techniques.

Paper	Acquisition Method	Test Fruit	Target Disease	Environment	Automation/ Manual	Coverage Area	Segmentation Method	Classifier	Accuracy
Jhuria [7]	Apple Image dataset	Apples	Apple scab and Apple rot	Off-Site Controlled Background and Lighting	Images Taken Manually	Single fruit	Color and Texture Extraction	Neural Network	High
Dubey [6]	Apple Image dataset	Apples	Apple blotch, rot, and scab	Off-Site Controlled Background and Lighting	Images Taken Manually	Single fruit	K-means segmentation	Support Vector Machine	High
Unay [18]	Monochrome Digital Camera with multiple bandpass filters	Jonagold Apples	Skin defects, bruises, and rot	Off-Site Controlled Background and Lighting	Images Taken Manually	Single fruit	Isodata Thresholding	Support Vector Machine	89.20%
Li [9]	2 Monochrome cameras with a bandpass filter	Apples	Surface defects	Off-Site Closed Chamber	Full automation	Multiple fruits	Simple subtraction thresholding	Neural Network	93%
Xing [22]	Multispectral Camera (400-1000 nm)	Golden Delicious Apples	Apple bruises	Off-Site Closed Chamber	Full automation	Multiple fruits	Simple thresholding	Principal Component Analysis	86%

Table 5.
Review of postharvest disease detection techniques.

For a reliable, rapid, and field-based disease detection system, a new preharvest automated method is required. A fusion of techniques such as VOC profiling and NIR imaging methods could be integrated into a robot for processing a large number of plants. Because of the uncertainty of lighting and other conditions of the field environment more advanced tools are required to capture data without being affected. An agriculture robot can be designed to move in agriculture fields to detect stresses in areas while providing position information. RGB and NIR imaging methods could be integrated into a robot and used in synchronization to measure overall plant health using the Normalized Differential Vegetative Index (NDVI). Because of its low cost, NIR imaging techniques can be very efficient and effective when integrated with a robot. The detection of stressful areas in a field with GPS information can be used for selective pesticide spraying.

Author details

Maki K. Habib* and Hashem Rizk
American University in Cairo, New Cairo, Egypt

*Address all correspondence to: maki@aucegypt.edu

References

[1] M. Jhuria, A. Kumar, and R. Borse, "Image processing for smart farming: Detection of disease and fruit grading," in 2013 IEEE Second International Conference on Image Information Processing (ICIIP-2013), Dec. 2013, pp. 521–526, doi: 10.1109/ICIIP.2013.6707647.

[2] S. R. Dubey and A. S. Jalal, "Detection and Classification of Apple Fruit Diseases Using Complete Local Binary Patterns," in 2012 Third International Conference on Computer and Communication Technology, 2012, pp. 346–351, doi: 10.1109/ICCCT.2012.76.

[3] D. Pimentel, R. Zuniga, and D. Morrison, "Update on the environmental and economic costs associated with alien-invasive species in the United States," Ecol. Econ., vol. 52, no. 3, pp. 273–288, Feb. 2005, doi: 10.1016/j.ecolecon.2004.10.002.

[4] S. Sankaran, A. Mishra, R. Ehsani, and C. Davis, "A review of advanced techniques for detecting plant diseases," Comput. Electron. Agric., vol. 72, no. 1, pp. 1–13, Jun. 2010, doi: 10.1016/j.compag.2010.02.007.

[5] M. López *et al.*, "Innovative tools for detection of plant pathogenic viruses and bacteria," Int. Microbiol. Off. J. Span. Soc. Microbiol., vol. 6, pp. 233–243, Jan. 2004, doi: 10.1007/s10123-003-0143-y.

[6] Q. Li, M. Wang, and W. Gu, "Computer vision based system for apple surface defect detection," Comput. Electron. Agric., vol. 36, no. 2, pp. 215–223, Nov. 2002, doi: 10.1016/S0168-1699(02)00093-5.

[7] S. Sannakki, V. Rajpurohit, V. Nargund, A. Kumar, and P. Yallur, "Leaf Disease Grading by Machine Vision and Fuzzy Logic," Int J, vol. 2, Nov. 2010.

[8] R. Oberti, M. Marchi, P. Tirelli, A. Calcante, M. Iriti, and E. Sciences-DiSAA, "The CROPS agricultural robot: application to selective spraying of grapevine's diseases," p. 8.

[9] B. Alberts, A. Johnson, J. Lewis, M. Raff, K. Roberts, and P. Walter, "General Principles of Cell Communication," Mol. Biol. Cell 4th Ed., 2002. Available: https://www.ncbi.nlm.nih.gov/books/NBK26813/ [Accessed: Nov. 09, 2020].

[10] J. J. Belasque, M. C. G. Gasparoto, and L. G. Marcassa, "Detection of mechanical and disease stresses in citrus plants by fluorescence spectroscopy," Appl. Opt., vol. 47, no. 11, pp. 1922–1926, Apr. 2008, doi: 10.1364/AO.47.001922.

[11] E. C. Lins, J. Belasque, and L. G. Marcassa, "Detection of citrus canker in citrus plants using laser induced fluorescence spectroscopy," Precis. Agric., vol. 10, no. 4, pp. 319–330, Aug. 2009, doi: 10.1007/s11119-009-9124-2.

[12] J. West, C. Bravo, R. Oberti, D. Lemaire, D. Moshou, and H. McCartney, "The Potential of Optical Canopy Measurement for Targeted Control of Field Crop Diseases," Annu. Rev. Phytopathol., vol. 41, pp. 593–614, Feb. 2003, doi: 10.1146/annurev.phyto.41.121702.103726.

[13] S. Lenk *et al.*, "Multispectral fluorescence and reflectance imaging at the leaf level and its possible applications," J. Exp. Bot., vol. 58, no. 4, pp. 807–814, Mar. 2007, doi: 10.1093/jxb/erl207.

[14] C. Bravo, D. Moshou, J. West, A. McCartney, and H. Ramon, "Early Disease Detection in Wheat Fields using Spectral Reflectance," Biosyst. Eng., vol. 84, no. 2, pp. 137–145, Feb. 2003, doi: 10.1016/S1537-5110(02)00269-6.

[15] A. Vallat, H. Gu, and S. Dorn, "How rainfall, relative humidity and temperature influence volatile emissions from apple trees in situ," Phytochemistry, vol. 66, no. 13, pp. 1540–1550, Jul. 2005, doi: 10.1016/j.phytochem.2005.04.038.

[16] J. M. Cevallos-Cevallos, R. Rouseff, and J. I. Reyes-De-Corcuera, "Untargeted metabolite analysis of healthy and Huanglongbing-infected orange leaves by CE-DAD," Electrophoresis, vol. 30, no. 7, pp. 1240–1247, Apr. 2009, doi: 10.1002/elps.200800594.

[17] S. B. Patil and D. S. K. Bodhe, "Leaf disease severity measurement using image processing," p. 6, 2011.

[18] G. Xu, F. Zhang, S. G. Shah, Y. Ye, and H. Mao, "Use of leaf color images to identify nitrogen and potassium deficient tomatoes," Pattern Recognit. Lett., vol. 32, no. 11, pp. 1584–1590, Aug. 2011, doi: 10.1016/j.patrec.2011.04.020.

[19] J. Xing, C. Bravo, P. T. Jancsók, H. Ramon, and J. De Baerdemaeker, "Detecting Bruises on 'Golden Delicious' Apples using Hyperspectral Imaging with Multiple Wavebands," Biosyst. Eng., vol. 90, no. 1, pp. 27–36, Jan. 2005, doi: 10.1016/j.biosystemseng.2004.08.002.

[20] A. Camargo and J. Smith, "Image pattern classification for the identification of disease causing agents in plants," Comput. Electron. Agric., vol. 66, pp. 121–125, May 2009, doi: 10.1016/j.compag.2009.01.003.

[21] T. Pobkrut and T. Kerdcharoen, "Soil sensing survey robots based on electronic nose," in 2014 14th International Conference on Control, Automation and Systems (ICCAS 2014), Oct. 2014, pp. 1604–1609, doi: 10.1109/ICCAS.2014.6987829.

[22] D. Unay and B. Gosselin, "Thresholding-based segmentation and apple grading by machine vision," Jan. 2005.

Chapter 5

Brain-Inspired Spiking Neural Networks

Khadeer Ahmed

Abstract

Brain is a very efficient computing system. It performs very complex tasks while occupying about 2 liters of volume and consuming very little energy. The computation tasks are performed by special cells in the brain called neurons. They compute using electrical pulses and exchange information between them through chemicals called neurotransmitters. With this as inspiration, there are several compute models which exist today trying to exploit the inherent efficiencies demonstrated by nature. The compute models representing spiking neural networks (SNNs) are biologically plausible, hence are used to study and understand the workings of brain and nervous system. More importantly, they are used to solve a wide variety of problems in the field of artificial intelligence (AI). They are uniquely suited to model temporal and spatio-temporal data paradigms. This chapter explores the fundamental concepts of SNNs, few of the popular neuron models, how the information is represented, learning methodologies, and state of the art platforms for implementing and evaluating SNNs along with a discussion on their applications and broader role in the field of AI and data networks.

Keywords: spiking neural networks, spike timing dependent plasticity, neuomorphic computing, artificial intelligence, low power, supervised learning, unsupervised learning, spatio-temporal learning, neuron models, spike encoding, winner take all, stigmergy

1. Introduction

Nature has provided innumerable examples of very efficient solutions to complex problems with seemingly simple rules. With these as inspiration, many engineering problems are tackled using bioinspired techniques. A few of bioinspired techniques are evolutionary and genetic algorithms, stigmergy, hidden Markov models, belief networks, neural networks, etc. These are applicable in a wide variety of domains from robotics [1], communication systems, routing [2], building construction [3], scheduling, optimization, machine intelligence, etc. The brain is a very efficient computing element capable of performing complex tasks. This is possible due to massively parallel computation being performed by the vast number of cells called neurons in the brain while consuming very little energy. This has inspired a domain of algorithms and techniques called artificial intelligence (AI) where machines are programmed to learn and then solve complex tasks. The recent advances in high performance computing and theoretical advances into statistical learning methodologies have enabled a widespread use of AI techniques for tasks such as pattern recognition, natural language understanding, speech recognition, computer vision, odor recognition, machine translation, medical diagnosis, gaming, autonomous driving, path planning, autonomous robots, financial market modeling and the list goes on. Solving these kinds of problems with efficiency is not possible with the traditional computing paradigms. These algorithms are mimicking biology or are inspired from biology to tackle the above problems. For example, it is not humanly possible to have traditional software program coded to classify an image of a simple object such as a cup with reasonable accuracy, considering the innumerable variations available in terms of shape, size, color, etc. However, this is a trivial task for a human being as our brains learn to identify the salient features of an object. The inner working of the brains, especially the way it processes information is the inspiration behind a class of AI techniques called neural networks.

AI requires a large amount of compute power while churning through massive amounts of data. Today's real-world tasks require different sets of AI models with different modalities to interact with each other, hence needing a large pipeline with complex data dependencies. Training is time-consuming, while needing efficient multi-accelerator parallelization. Even with such advances we are nowhere close to the compute power or the efficiency of a human brain. Human brain is still a mystery and is a very actively researched topic. Several neuron models are proposed to mimic various aspects of how the brain works with the limited understand we have up till now.

Spiking neural networks (SNNs) are networks made up of interconnected computing elements called neurons. SNNs try to mimic biology to incorporate the efficiencies found in nature. These neurons use spikes to communicate with each other. SNNs are third generation of neural networks [4] and are gaining popularity due to its potential for very low energy dissipation due to their event-driven and asynchronous operation. SNNs are also interesting because of their ability learn in a distributed way using a technique called Spike Timing Dependent Plasticity (STDP) learning [5]. STDP relies on sparsely encoded spiking information among local neurons. SNNs are capable of learning rich spatio-temporal information [6]. In principle, SNNs can be fault tolerant due to its ability to re-learn and adapt the connections with other neurons, akin to how the brains learn. Also SNNs can natively interface with specialized hardware sensors which mimic biological vision (Dynamic Vision Sensor) and hearing (Dynamic Audio Sensor) [7] as they directly transduce sensory information to spikes.

In the rest of the chapter, a brief introduction on neuron biology and artificial neuron models is presented, followed by discussion on information representation as spikes, different learning methodologies, tools, and platforms available for simulating and implementing SNNs and finally few case studies as examples of SNN usage.

2. Neuron models

In this section, a brief overview of the biological neuron processes is provided to understand the inference and learning dynamics of SNNs. A few popular neuron models are discussed at a high level to make the reader aware of the diversity of such research and its use in SNNs.

2.1 Biological neuron

Complex living organisms have specialized cells called neurons, which are the fundamental unit of central nervous system. Neurons can transmit and receive signals in the form of electrical impulses. In a human brain, there are an estimated 200 billion neurons. Also, there are several different types of neurons in the body. In general, a neuron consists of a cell body or soma consisting of cell machinery, nucleus, dendrites, and an axon as shown in **Figure 1**.

The dendrites receive information from other neurons, and this causes a voltage buildup on the cell body. When this membrane potential reaches a certain threshold, an electrical impulse is generated, and the axon transmits this spike away from the cell body to other neurons. After a spike is generated, the neuron returns to a lower potential called resting potential. Also, immediately after a spike is generated, the neuron cannot generate another spike for a short duration called the refractory period. The axon terminates at axon terminals which interface with dendrites of other neurons; this is called as a synapse. A synapse is connection between a pre synaptic neuron (which generates electrical impulse) and a postsynaptic neuron (receives the spike information) as shown in **Figure 1**. The synapse is not a direct connection, instead it consists of a gap called synaptic cleft as shown in **Figure 2**. Discussion about astrocyte cells is presented later in Section 4.5.

When an electrical impulse reaches the synapse, the presynaptic neuron releases certain chemicals called neurotransmitters into the synaptic cleft. The postsynaptic neuron picks up these neurotransmitters eventually causing the postsynaptic neurons membrane potential to either increase or decrease. The brain learns by strengthening or weakening the existing synaptic connections or by making new synaptic connections or dissolving those which are no longer needed. In this way, the synapses make the brain plastic and provide the ability to learn. Also, the strength of the synapse also matters for learning as it can modulate the amount of neurotransmitters released in the synaptic cleft resulting in a stronger or weaker synapse and depending on the type

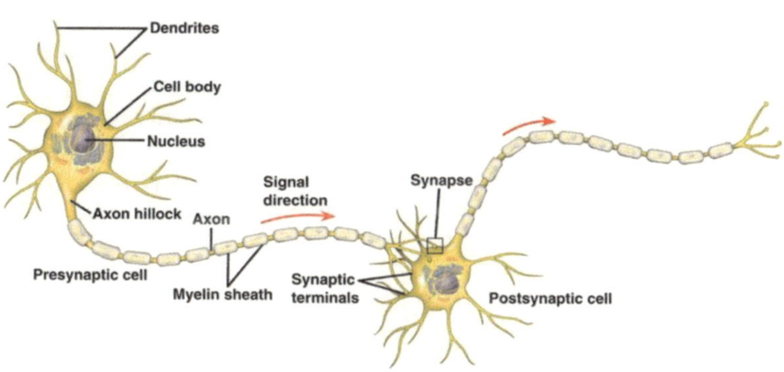

Figure 1.
Neurons (by unknown author, licensed under CC BY-SA https://creativecommons.Org/licenses/by-sa/3.0/).

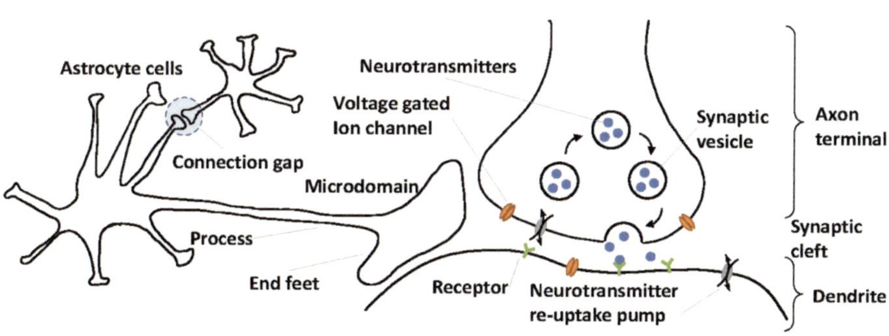

Figure 2.
Neuronal synapse along with astrocyte cells (author created).

of neurotransmitters released, the synapse can be excitatory or inhibitory. An excitatory synapse is one which would increase the membrane potential of the post synaptic neuron; conversely, an inhibitory synapse would decrease the membrane potential. Based on these fundamental concepts, several researchers have proposed various neuron models over the decades. We do not yet fully understand the inner workings of brains and is still an active field of research. New neuron models are being proposed frequently as our understanding of biology increases. A few neuron models are listed below, followed by an overview of select models.

2.2 Artificial neuron models

Some of the models proposed try to mimic biology for the purpose of understanding and modeling neuro-physiological processes and some models more oriented toward computing purposes. A few of neuron models to consider are McCulloch and Pitts [8], Hodgkin-Huxley [9], Perceptron [10], Izhikevich [11] Integrate and fire [12], Leaky integrate-and-fire [13], Quadratic integrate-and-fire [14], Exponential integrate-and-fire [15], Generalized integrate-and-fire [16], Time-varying integrate-and-fire model [17], Integrate-and-fire or burst [18], Resonate-and-fire [19], and Bayesian neuron model [20].

2.3 Hodgkin and Huxley neuron model

Hodgkin and Huxley [9] studied the giant axon of the squid and found currents induced by different types of ions namely sodium ions, potassium ions, and leakage current due to calcium ions. The cell consists of voltage-dependent ion channels which regulate the concentration of these ions across the cell membrane. For the sake of simplicity, at a high level, the total membrane current is the sum of current induced by membrane capacitance and the ion channel currents as shown in Eq. (1), where I_i is the ionic current density, V is the membrane potential, C_M is the membrane capacitence per unit area, t is time, and I_{Na}, I_K, I_l are the sodium, potasium, and leakage current induced by calcium and other ions.

$$I = I_c + I_i \tag{1}$$

$$I_c = C_M \frac{\mathrm{dV}}{\mathrm{d}t} \tag{2}$$

$$I_i = I_{Na} + I_K + I_l \tag{3}$$

They also describe gating variables to control the ion channels and the resting potential of the cell. When the membrane potential increases significantly above the resting potential, the gating variable activates and then deactivates the channels resulting in a spike. This is a very simplified model and has several limitations [21].

2.4 Izhikevich neuron model

Izhikevich neuron model [11] is more biologically plausible as shown in equations below.

$$\begin{aligned} v' &= 0.04v^2 + 5v + 140 - u + I \\ u' &= a(bv - u) \\ \mathit{If}\ v &\geq 30mV, \mathit{then} \begin{cases} v = c \\ u = u + d \end{cases} \end{aligned} \tag{4}$$

Where v is the membrane potential, u is the recovery variable, I is the current, and a, b, c and d are neuron parameters. Various biologically plausible firing patterns can be modeled using this model as shown in **Figure 3**.

Over time, if a biological neuron does not spike, then any potential builtup would dissipate. This phenomenon is modeled by several variations of Leaky Integrate and Fire (LIF) models. LIF neuron model is very popular due to its ease of implementation as a software model and for developing dedicated hardware models. Digital hardware implementation is more popular than the analog variants, again due to its simplicity of design, fabrication, and scalability.

2.5 Discrete leaky integrate and fire

A typical generic LIF model adapted for discrete implementation [22] is represented as:

Synaptic integration

$$V(t) = V(t-1) + \sum_{i=0}^{N-1} x_i(t) s_i \tag{5}$$

Leak integration

$$V(t) = V(t) - \lambda \tag{6}$$

Threshold, fire and reset

$$\text{If } V(t) \geq \alpha \text{ then Spike and } V(t) = R \tag{7}$$

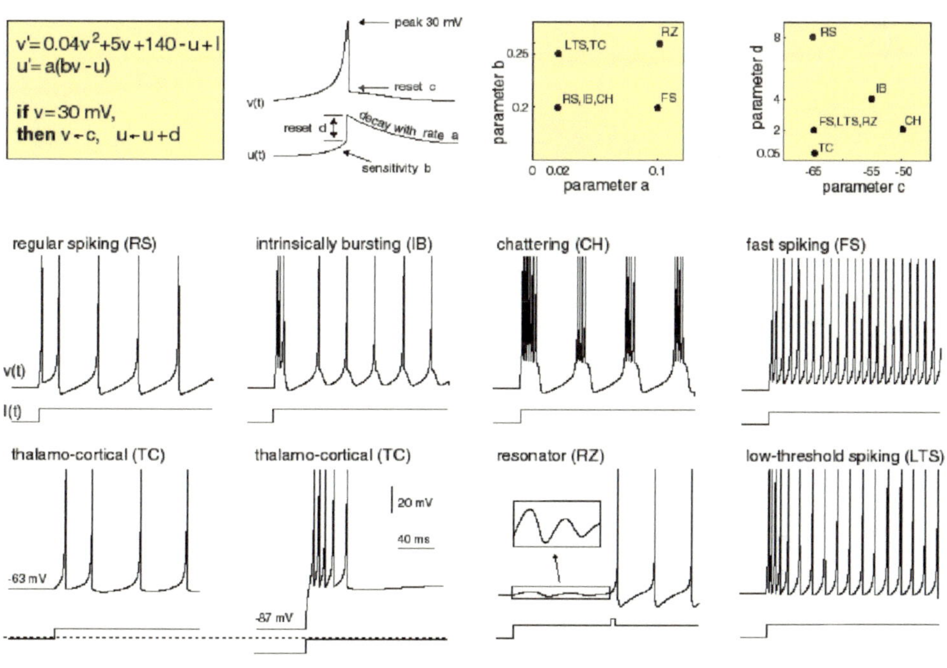

Figure 3.
Izhikevich neuron model [11].

Where $V(t)$ is the membrane potential, t is discrete time step, N is the number of synapses, $x_i(t)$ is the ith synapse, s_i synaptic weight of ith synapse, λ is leak, α is spiking threshold, and R is the resting potential. A spike value is 1, otherwise 0. Whenever, a spike occurs on a synapse $x(t)$ then the synaptic weight gets accumulated increasing the membrane potential. Every time step a leak is applied and finally when the membrane potential reaches a threshold α, the neuron spikes and the membrane potential is reset to a resting value R.

2.6 Bayesian neuron model

Bayesian neuron (BN) model is proposed in [20]. BN model is a stochastic neuron model. When the membrane potential reaches the threshold a BN model fires a spike stochastically. It generates a spike based on a Poisson process where neuron Z fires at time t with a probability proportional to its membrane potential at time t. The membrane potential $u(t)$ is computed as:

$$u(t) = w_0 + \sum\nolimits_{i=1}^{n} w_i y_i(t) \tag{8}$$

Where the weight of the synapse between i^{th} presynaptic neuron y_i and Z is w_i. If y_i fires a spike at time t, then $y_i(t)$ is 1. The intrinsic excitability is w_0. The firing probability of this stochastic neuron model depends exponentially on the membrane potential $u(t)$ as:

$$probability(Z\, fires\ at\ time\ t) \propto \exp\left(u(t)\right) \tag{9}$$

To generate a Poisson process with time-varying rate $\lambda(t)$, the *Time-Rescaling Theorem* is used. According to this theorem, when spike arrival times v_k follow a Poisson process of instantaneous rate $\lambda(t)$, the time-scaled random variable $\Lambda_k = \int_0^{v_k} \lambda(v)dv$ follows a homogeneous Poisson process with unit rate. Then the interarrival time τ_k satisfies exponential distribution with unit rate.

$$\tau_k = \Lambda_k - \Lambda_{k-1} = \int_{v_{k-1}}^{v_k} \lambda(v)dv \tag{10}$$

τ_k represents a generated random variable satisfying an exponential distribution with unit rate. v_k is the next time to spike. As shown in Eq. (10), the instantaneous rates from Eq. (8) is cumulated until the integral values is greater than or equal to τ_k. At this point of time, a spike is generated as it implies that the interspike interval has passed. Poisson spiking behavior is achieved in this way reflecting the state of the neuron. Other stochastic neuron behaviors can be easily constructed by stochastically varying different parameters of the model.

3. Information representation

SNNs understand the language of spikes, and it is necessary to decide what is the best possible way to represent real-world data to achieve best possible training of the network and efficient inference. Different coding techniques model different aspects of input spectrum. Some of the spike coding techniques are described below to get an intuition of signal representation using spikes.

3.1 Rate coding

With rate coded spike trains, the information is encoded in the number of spikes over a specified temporal window. The firing rate ν_k, over k trials is shown in Eq. (11), where n_k^{sp} is the number of spikes over k trials over a temporal window T and is the number of trials [23].

$$\nu_k = \frac{n_k^{sp}}{T} \tag{11}$$

Evidence of rate coding is experimentally shown in sensory and motor systems [24]. The number of spikes emitted by the receptor neuron increases with the force applied to the muscle.

If the rate ν is defined via a spike count over a temporal window of duration T, the exact firing time of a spike does not matter [23]. We can define it as a Poisson process where spikes events are stochastic and independent of each other with an instantaneous firing rate ν. In a homogeneous Poisson process, the probability to find a spike in a short interval Δt is

$$P_F(t; t + \Delta t) = \nu \Delta t \tag{12}$$

Therefore, the instantaneous firing rate is

$$\nu = \lim_{\Delta t \to 0} \frac{P_F(t; t + \Delta t)}{\Delta t} \tag{13}$$

The expected number of spikes for the temporal window T is

$$\langle n^{sp} \rangle = \nu T \tag{14}$$

To summarize, the experimental procedure of counting spikes over a time T and dividing by T gives an empirical estimate of the rate ν of the Poisson process. When recording an experiment over several trials, the spike response can be represented via a Peri-Stimulus-Time Histogram (PSTH) with bin width Δt as shown in **Figure 4**. The number of spikes $n_k(t; t + \Delta t)$ summed over all repetitions K of the experiment is a measure of the typical activity of the neuron between time t and $t + \Delta t$. Therefore, the spike density can be represented as shown in Eq. (15).

$$\rho = \frac{1}{\Delta t} \frac{n_k(t; t + \Delta t)}{K} \tag{15}$$

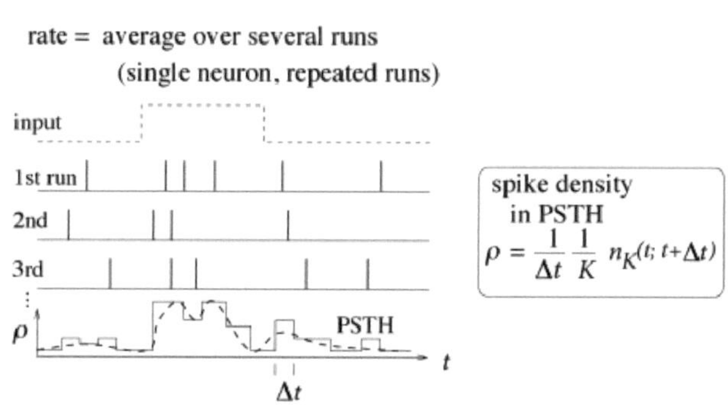

Figure 4.
The Peri-stimulus-time histogram and the average time-dependent firing rate [23].

A spike train $S(t)$ is a sum of δ functions with a spike occurring at t_s.

$$S(t) = \sum_s \delta(t - t_s) \tag{16}$$

The instantaneous firing rate is the expectation over trials.

$$v(t) = \langle s(t) \rangle \tag{17}$$

An empirical estimate of the instantaneous firing rate can be deduced as shown in Eq. (18). It implies that the PSTH as described above represents the instantaneous firing rate.

$$\nu(t) = \frac{1}{K\Delta t} \sum_{k=1}^{K} n_k^{sp}(t) \tag{18}$$

The average firing rate can be computed for a single neuron, or for a population of neurons representing a class over a single run or over several trials. Rate coding over a time window is suitable for representing the strength of stimulation. On the other hand, population-based rate coding could convey the same information by employing several neurons in a shorter temporal window. The latter trades quick response over a number of neurons. There is evidence of Purkinje neurons demonstrating information coding which is not just firing rate but also the timing and duration of nonfiring, quiescent periods [25, 26].

3.2 Temporal coding

If the time of spike occurrence in a temporal window carries information, then such coding is referred to as temporal coding. In such coding schemes the quiescent periods and the spiking time both carry information. There are several evidences in biology demonstrating this behavior [27, 28]. A typical temporal code is shown in **Figure 5A**, where the time interval of spike to start of stimulus caries information. These are sometimes referred to as pulse codes. Another variation is Rank Order Coding, which uses the relative timing of spikes across a population of cells. Rank order codes look at time to spike across the neuron population and a rank order can

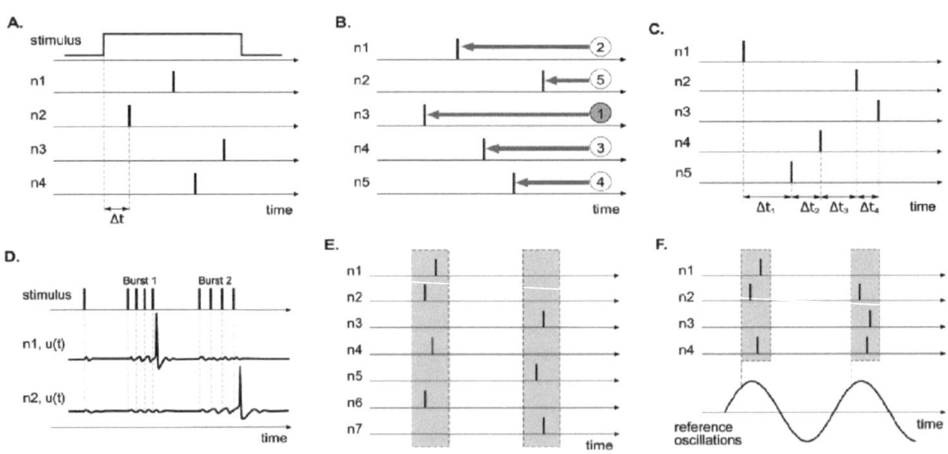

Figure 5.
Different strategies for information coding with spikes (refer to [29] for details). (A) Time to first spike coding (B) rank order coding (C) latency coding (D) resonant burst coding (E) synchrony coding (F) phase coding.

be implied from the firing order among the neurons in the population as described in **Figure 5B**.

There is evidence suggesting that simple temporal averaging of firing rate is too simplistic to model neuronal circuits in the brain [30]. To address some of the shortcomings, several derivations of coding schemes based on different combinations of above concepts are widely used. Few of the common schemes and some task specific coding schemes are Rate code, Time to spike code, Time-to-first-spike: Latency code [31], Reverse time to spike code, Weighted spike code [32], Burst code [33], Population code, Population rate, Rank order code [34], Phase-of-firing code [35, 36], Place code [37], etc. **Figure 5** summarizes a few coding strategies. These coding schemes require appropriate algorithms for converting real-world data to spikes and vice versa. A few common conversion techniques are discussed in the next section.

3.3 Spike transduction

SNNs understand the language of spikes; therefore, we must transform the real-world data to appropriate spike representation and subsequently transform the output spikes to real-world formats for human consumption. There are several encoding and decoding algorithms available to achieve this goal. Several heuristics are also employed. Some of the coding techniques mentioned above infer a specific coding/ decoding scheme. Based on the nature of application (such as images, audio, video, financial data, user activity data), one must choose which is the best approach.

Image pixel values are binned and proportional firing rates are assigned to different neurons in the receptive fields for each pixel neuron, hence generating random process with rate coding [38]. Since spikes have no polarity positive and negative spike, subchannels can be used to represent richer encoding of data. In threshold-based schemes, a spike is generated when the input signal intensity crosses a threshold. Real numbers are compared against different thresholds, and positive and negative spikes are produced accordingly which are rate coded [39]. BSA algorithm for encoding and decoding [40] is used for modeling brain-machine interfaces and neurological processes in the brain. The work presented by the authors of [41] provides details on step-forward (SF), and moving-window (MW) encoding schemes. In SF scheme, a baseline $B(t)$ intensity for the input signal is set and a positive spike is generated if the intensity is above the baseline by the threshold $B(t) + Th$ amount and the baseline is updated to this new value $B(t) = B(t-1) + Th$. Conversely a negative spike is generated if the signal intensity is below $B(t) - Th$ and the new baseline is adjusted as $B(t) = B(t-1) - Th$. MW scheme is like SF scheme except that the baseline is set based on the mean of signal intensities. These schemes are suitable for encoding continuous value signals. The above examples are only a limited set of algorithms out of a vast majority of methods to convert diverse signal formats to spikes.

4. Learning principles for SNN

Hebb postulated that synaptic efficacy increases from a presynaptic neuron if it repeatedly assists the post synaptic neuron [42]. This forms the fundamentals of STDP rule for learning. STDP mimics biology where a synapse is strengthened when a presynaptic spike occurs before a post synaptic spike in close intervals, this is called Long-Term Potentiation (LTP). On the other hand, the synapse is weakened if the post synaptic neuron fires before the presynaptic neuron in close intervals. This is called as Long-Term Depression (LTD). In biology neurons are highly selective due to lateral inhibition. This allows for them to learn discriminatory and unique features in an unsupervised manner leading to an emergent Winner Take All (WTA) behavior. Apart from this the biological system demonstrates homeostasis to maintain overall stability. These are key principles in SNN modeling. There are several ways to achieve WTA and homeostasis behavior, some directly modify the neuron state, others use neural circuits. One such example with a scalable neural circuit [43] is shown in **Figure 6**. A WTA network consists of inhibitor neurons suppressing the activation of other lateral symbol neurons as shown in **Figure 6(a)**. To assist in homeostasis a normalization of the excitations of one neural circuit compared to others can be achieved using a Normalized Winner Take All (NWTA) network as shown in **Figure 6(b)**. Where an upper limit (UL) neuron uniformly inhibits all symbol neurons if they are firing beyond a desirable high threshold. On the contrary if the symbol neurons are firing below a desired low threshold, then the lower limit (LL) neuron triggers an excitor (Ex) neuron to uniformly boost the firing rate of all symbol neurons. In this manner all independent neural circuits within an SNN fire in the dynamic range of excitations of the overall network. Both hard and soft WTA behavior can be achieved based on the amount of inhibition generated. In Hard WTA only one symbol neuron is active whereas in soft WTA more than one symbol neuron is active providing richer context.

SNNs can learn in both unsupervised and supervised modes. WTA concepts are essential part of unsupervised learning as the neuron with highest excitation inhibits the lateral neurons the strongest hence enabling it to preferentially pick up unique features. Unsupervised learning is possible by employing a teacher signal which excites the specific neurons to fire thereby allowing it to learn the features represented by the input signal. STDP based learning has its advantages of being able to model spatio-tempotal dynamics. Where the spatial component refers to localized activity/learning and temporal component refers to additional information representation by the spike intervals along the time axis. With the constant advances in SNN research, native STDP based rules are catching up to the more popular backpropagation-based learning methods used in Artificial Neural Networks (ANN). STDP lends itself for efficient localized and distributed learning, which is a huge advantage over other learning methods. Also SNNs can be adapted to model memories in the form of Long Short-Term Memory networks [39] which shows that recurrent learning behavior is also possible. The following sub-sections discus few learning rules used in training SNNs along with a brief introduced to backpropagation-based learning.

4.1 Classic STDP rule

A classic STDP rule [44] is shown in **Figure 7**. The STDP curve tries to approximate experimentally observed behavior.

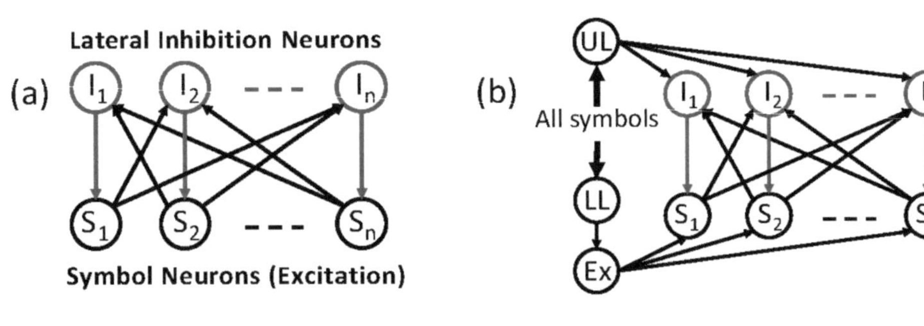

Figure 6.
(a) Winner take all network (b) normalized winner take all network [43].

Here ΔW is the weight update plotted against $\Delta t = t_{pre} - t_{post}$ representing the interval between the presynaptic and post synaptic spike. This approximation is represented in Eq. (19)

$$\Delta W = \begin{cases} a^+ \exp\left(\frac{t_{pre} - t_{post}}{\tau^+}\right) if\ t_{pre} \le t_{post}\ (LTP) \\ \\ -a^- \exp\left(-\frac{t_{pre} - t_{post}}{\tau^-}\right) if\ t_{pre} > t_{post}\ (LTD) \end{cases} \tag{19}$$

Where a^+, a^- are the learning rates and τ^+, τ^- are the time constants for LTP and LTD, respectively. There are several variations of the STDP curves available in the literature and the reader is encouraged to explore this topic further.

4.2 Simplified stable STDP rule with efficient hardware model

There are two broad categorizations of STDP rules, additive and multiplicative STDP [38]. Multiplicative rule tends to be more stable than additive rule. In additive rules the weight changes are independent of current weight and requires additional constraints to keep the values in operating bounds. These weight changes however produce bimodal distribution resulting in strong competition. In multiplicative rule presented in [38], the weight change is inversely proportional to the current weight making it inherently stable and resulting in a unimodal distribution. This distribution lacks synaptic competition which is desirable for learning discriminative features. For such rules, competition must be introduced in a different method. The stable multiplicative rule is further explored below and simplified for efficient implementation. Here the STDP rule is modeled such that weight change of a synapse has an exponential dependence on its current weight as shown in **Figure 8** (a). Update for the weight w_i of ith synapse of the neuron is calculated as below.

If

$$t_{post} - t_{pre} < \tau_{LTP} \tag{20}$$

then,

$$\Delta w_i = \eta_{LTP} e^{-w_i}, w_i = w_i + \Delta w_i$$

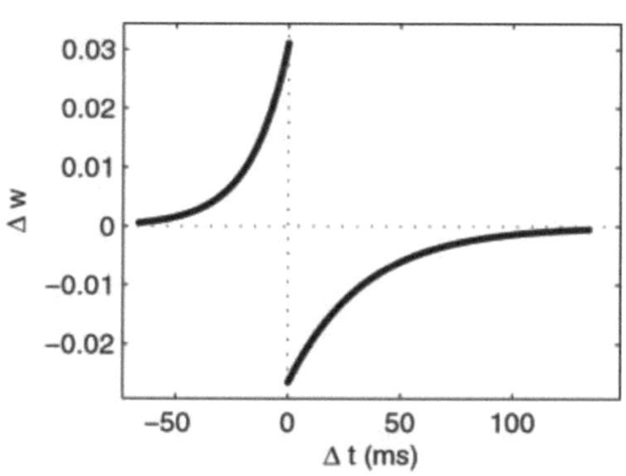

Figure 7.
Classic STDP curve [44].

If

$$t_{post} - t_{pre} > \tau_{LTP} \text{ or } t_{pre} - t_{post} < \tau_{LTD}$$

then,

$$\Delta \mathrm{w_i} = \eta_{\mathrm{LTD}} e^{\mathrm{w_i}}, \mathrm{w_i} = \mathrm{w_i} - \Delta \mathrm{w_i} \tag{21}$$

Where t_{post} and t_{pre} are the pre and post-synaptic neuron spiking time steps, τ_{LTP} and τ_{LTD} are the LTP and LTD window and η_{LTP} and η_{LTD} are the LTP and LTD learning rates respectively. Plasticity is implemented with LTP and LTD windows as shown in **Figure 8** (b). This rule is called as Exp rule.

The Exp STDP rule requires an exponential and a multiplication operation for both LTP and LTD for each synapse. From the perspective of efficient digital hardware implementation these are expensive operations in terms of circuit area and computation time. Quantized 2-power shift rule (Q2PS), which approximates the Exp rule in Eq. (20) and Eq. (21) by removing both multiplication and exponential. The approximation is summarized in Eq. (22) and Eq. (23).

If

$$t_{post} - t_{pre} < \tau_{LTP}$$

$$\Delta w_i = \eta_{LTP} 2^{-w_i} = 2^{\eta'_{LTP} - w_i} \tag{22}$$

If

$$t_{post} - t_{pre} > \tau_{LTP} \text{ or } t_{pre} - t_{post} < \tau_{LTD}$$

$$\Delta w_i = \eta_{LTD} 2^{w_i} = 2^{\eta'_{LTD} + w_i} \tag{23}$$

where $\eta'_{LTP} = \log_2 \eta_{LTP}$ and $\eta'_{LTD} = \log_2 \eta_{LTD}$. Let $Q = \eta'_{LTP} - w_i$ for LTP and $Q = \eta'_{LTD} + w_i$ for LTD. Let $\overline{Q}$ be the quantization of Q through priority encoding. Priority encoding compresses a binary representation of a number to value with only the most significant bit being active as rest of the active bits have no priority. For example, the binary representation of $Q = 13$ is 1101 and the priority encoded value is 1000, hence $\overline{Q} = 8$. Based on this quantization method, the synaptic weight change can be easily computed by left shifting 1 by $\overline{Q}$ or right shifting if negative as shown in Eq. (24).

$$\Delta w_i = \begin{cases} 1 \ll |\overline{Q}|, if\ \overline{Q} > 0 \\ 1 \gg |\overline{Q}|, if\ \overline{Q} < 0 \end{cases} \tag{24}$$

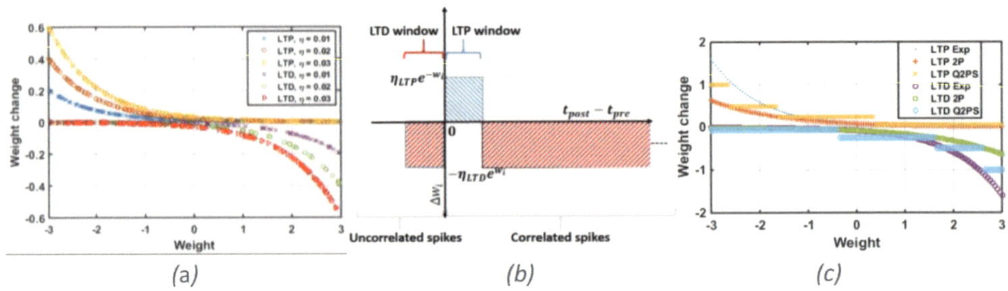

Figure 8.
(a) Current weight vs weight change for learning rates (b) STDP windows (c) Comparison of Exp, 2P and Q2PS STDP rules [38].

where $\ll$ and $\gg$ represent binary shift left and shift right operations, respectively. This approximation allows implementation of the STDP rule presented in Eq. (20) and Eq. (21) on digital hardware by using a priority encoder, negligibly small lookup to determine $|\overline{Q}|$ from the encoded value, barrel shifter and an adder circuit. Please note that, based on Eq. (22) and Eq. (23), Δw_i should be calculated as $2^{\overline{Q}}$, which can be obtained by shifting value 1 by $\overline{Q}$. **Figure 8** (c) compares the Δw_i calculated using the Exp, 2P and Q2PS rules, with a learning rate of 0.08 for all the cases. Here 2P rule is same as Q2PS rule except that 2 is raised to the power of Q. As we can see, the Q2PS rule provides multi-level quantization, which enables similar quality of trained weights even with approximations when compared to Exp rule.

4.3 Overview of learning in artificial neural networks

With the tremendous advances in the field of ANNs, a growing body of research is available on various statistical learning algorithms. ANNs are inspired by biology but they do not mimic it. ANNs are made up of artificial neuron models specifically tuned for compute purposes and model a biological neuron at a very abstract level. An artificial neuron computes weighted sum of input signals and then an activation function computes the neuron output. In these networks' neurons transmit signals as real numbers. ANNs compute inference by transmitting the neuron signals in the forward direction. The learning happens usually via a method called Backpropagation. This algorithm computes the gradients based on the error signal produced by a cost function and propagates it back for each layer of neurons in the neural network. The weight updates are usually made using gradient descent algorithms. There are many flavors of gradient descent algorithms available in the literature. For back propagation to work the activation function must be differentiable. Unlike SNNs, where a spike is not differentiable. In general, ANNs have proven to be very effective in tackling a wide variety of problems. Using these algorithms as inspiration several modified STDP rules have been researched, one among them is discussed below. This overview is a very high-level introduction to some of the terminology required to understand the following section. The reader is encouraged to explore further on this topic.

4.4 Backpropagation-STDP

The Backpropagation-STDP (BP STDP) [45] algorithm uses the number of spikes in a spike trains as an approximation for the real value of an artificial neurons excitation. They also divide the time interval into sub-intervals such that each sub-interval contains zero or one spike.

In supervised training, the weight adjustment is governed by the STDP model shown in Eq. (25) and Eq. (26), in conjunction with a teacher signal. The teacher signal when applied to target neurons undergo weight change based on STDP and non-target neurons undergo weight changes based on anti-STDP. Anti-STDP is the opposite of STDP where LTP and LTD equations are swapped. Target neurons are identified by spike trains with maximum spike frequency (β) and non-target neurons are silent. The expected output spike trains z, are tagged with their input labels. Eq. (25) represents the weight change for a desired spike pattern $z_i(t)$ for the output layer neurons.

$$\Delta w_{ih}(t) = \mu \xi_i(t) \sum_{t'=t-\epsilon}^{t} s_h(t') \tag{25}$$

$$\xi_i(t) = \begin{cases} 1, z_i(t) = 1, r_i \neq 1 \; [t-\epsilon, t] \\ -1, z_i(t) = 0, r_i = 1 \; [t-\epsilon, t] \\ 0, otherwise \end{cases} \tag{26}$$

A target neuron would generate a spike $z_i(t) = 1$ and non-target neurons would remain silent $z_i(t) = 0$. Based on the expected output spike train target neuron should fire within the short STDP window $[t - \epsilon, t]$. Based on the presynaptic activity usually zero or one spike in the STDP window, the synaptic weights are increased proportionally. The presynaptic activity is the count of spikes in the $[t - \epsilon, t]$ interval denoted as $\sum_{t'=t-\epsilon}^{t} s_h(t')$. On the other hand, the non-target neurons upon firing undergo weight depression in the same way. The difference between the desired spike pattern and output spike pattern is used as the guide for identifying target neurons and non-target neurons as the backpropagation rule. Same methodology is used for each layer while back propagating. Among the several learning methods inspired by ANN algorithms a few use strategies where the ANN is trained in its native form and tuned based on a shadow SNN and finally use those adapted weights on SNN for inference.

4.5 Stigmergy assisted learning

Stigmergy is a methodology where several independent agents produce an emergent behavior through indirect interaction among themselves. This is facilitated with the help of asynchronous communication through traces left in the environment by individual agents. Stigmergy has been observed in nature and widely researched upon especially in insect colonies, these principles have been applied towards solving various engineering problems. Recent advances in neuroscience have shown evidence of another type of cells called astrocytes working in tandem with neurons to regulate the behavior of the central nervous system [46]. Astrocytes are star shaped cells with several branches called as processes. The end of these branches called as end feet interface with a synapse by wrapping around it creating a region around the synaptic cleft called as microdomain as shown in **Figure 2**. Astrocytes also interface the neurons apart from the synapse providing a closed loop feedback mechanism. They also interface with other astrocytes like a synapse, instead this is called as a gap junction. Gap junction facilitates communication between astrocyte cells only through chemical means. Astrocytes are functionally very diverse and play a very important role, only a high-level concept with limited detail is introduced for understanding of relevant discussion. With the help of calcium ions as a signaling mechanism along with the help of neurotransmitters the astrocytes help regulate the efficiency of synaptic transmission. These cells play a critical role in maintaining homeostasis, modulating LTP, LTD and structural plasticity in the brain.

Spiking activity results in release of neurotransmitters and change in concentration among different ions in the microdomain and extra cellular space. These

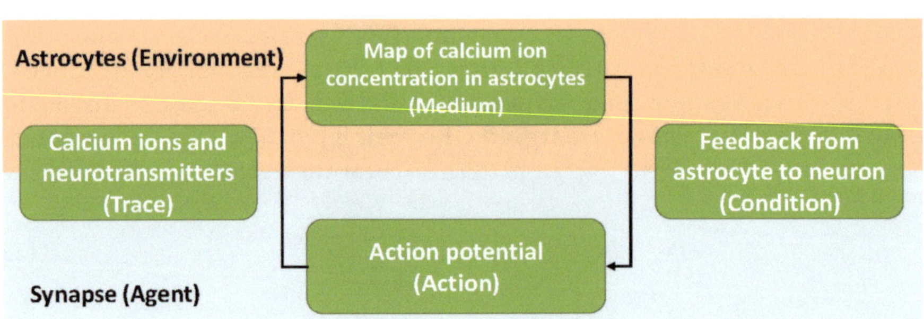

Figure 9.
Stigmergic interactions between astrocytes and neurons (modified from [47]).

changes are monitored as traces for indirect communication by astrocytes. Astrocytes themselves behaving like an environment with calcium ion concentration gradients within the cell acting as a medium for other neuron agents to indirectly infer these changes. This interaction creates a feedback mechanism in an asynchronous and distributed manner [47]. **Figure 9** shows the emergent stigmergy pattern in the brain. Short term activity and long-term activity gets communicated over a distance to other synapses over a spatial domain. Greater the distance, lower would be the influence. The details about the stigmergy based brain plasticity is presented in [47], interested readers are encouraged to explore further. This is a relatively new discovery and extensive research is underway to understand the role of astrocytes in overall brain mechanics.

5. SNN simulation tools and hardware accelerators

There are several spiking neural network simulation tools available which support biologically realistic neuron models for large scale networks. Some of the popular ones are:

Brian [48], is a free, open source simulator for spiking neural networks. This simulator is capable of running on several different platforms and is implemented in python making it extendable and easy to use.

NEST [49] is another simulator focusing on the dynamics, size and structure of neural systems both large and small. This tool is not intended for modeling the intricate biological details of a neuron.

NEURON [50] is simulation environment best suited for modeling individual neurons and their networks. This is popular among neuroscientists for its ability to handle complex models in a computationally efficient manner. Unlike above simulator, NEURON can handle morphological details of a neuron and is used to validate theoretical models with experimental data.

The above tools are commonly used in modeling biologically realistic neuron modes. They have their own unique interfaces and low-level semantics. An effort is made to smooth things out with a tool independent API package developed on Python programming language called PyNN [51]. The PyNN framework provides API support to model SNNs at a high level of abstraction of all aspects of neuron modeling and SNN representation, including populations of neurons, connections, layers etc. Though this provides high level abstraction, it also provides the ability to program at a low level such as adjusting individual parameters at the neuron and synapse level. To make things easy PyNN provides a set of library implementation for neurons, synapses, STDP models etc. They also provide easy interfaces to model various connectivity patterns among neurons like; all-to-all, small-world, random distance-dependent etc. These APIs are simulator independent making the code portable across different supported simulation tools and neuromorphic hardware platforms. It is relatively straightforward to add support to any custom simulation tool. PyNN officially supports BRIAN, NEST and NEURON SNN simulation tools. It is also supported on SpiNNaker [52] and BrainScaleS-2 [53] neuromorphic hardware systems. There are several more simulation tools which work with PyNN.

Cypress [54] is a C++ based SNN Simulation tool. This provides a C++ wrapper around PyNN APIs. Hence, extending the multi-platform reach of Cypress using C++ interface. It is also capable of executing networks remotely on neuromorphic compute platforms.

The BrainScaleS-2 [53] is a mixed-signal accelerated neuromorphic system with analog neural core, digital connectivity along with embedded SIMD microprocessor. It is efficient for emulations of neurons, synapses, plasticity models etc. This hardware based system is capable of evaluating models up to ten thousand times faster than real time.

The SpiNNaker [52] is another neuromorphic system custom built with digital multicore ARM processors. The SpiNNaker system (NM-MC-1) consists of custom chips each with eighteen cores sharing a local 128 MB RAM. The overall system scales to more than a million cores.

Apart from the above tools and platforms the are many custom SNN tools available to model SNNs easily for machine learning purposes. ANNarchy (Artificial Neural Networks architect) [55] is a custom simulator for evaluating SNNs. This is implemented in C++ language, along with acceleration support provided using OpenMP/CUDA. The network definitions are provided using python interface.

NeuCube [6] is a development environment for creation of Brain-Like Artificial Intelligence. The computational architecture is suited for modeling SNN applications across several domain areas. This tool supports the latest neural network models for AI purpose. It supports PyNN interface, hence extending its versatility. This tool can run on CPU, GPU and SpiNNaker platforms, also a cloud version of the tool is available.

TrueNorth [56] is another neuromorphic platform capable of evaluating SNNs at faster than real time and at very low power. They demonstrate running state of the art neural networks on the hardware platform scaling up to 64 million neurons and 16 billion synapses while the system consumes only 70 W of power out of which only 15 W is consumed by the neuromorphic hardware components. The hardware supports inference only, with learning performed off chip.

Loihi [57] is the latest offering in the neuromorphic SNN hardware. This hardware approach gets rid of crossbar architecture, which is prevalent in most previous neuromorphic implementations, lending itself to greater amount of flexibility. Loihi is also capable of on-chip learning which is a huge advantage in terms of online learning of synapses.

Other simulators capable of modeling software based models and models for custom neuromorphic hardware are presented in [20, 58–60]. This is still an ongoing field of research and there are several more accelerator-based simulators available hence the reader is encouraged to explore further. Neuromorphic hardware using more exotic hardware devices like memristors and phase change memories are also an active area of research, they are yet to make it to mainstream consumption hence they are only mentioned here.

6. Case studies

In this section few case studies are presented to bolster the concepts discussed in this chapter. The topics covered here include STDP learning dynamics, probabilistic graphical models as SNNs, SNN with BP-STDP based learning and SNNs on Neuromorphic Hardware.

6.1 STDP learning dynamics

A SNN is trained [38] to classify handwritten digits from the MNIST dataset using the STDP based learning rules Exp, Q2PS and 2P presented in Section 4.2. The authors build a three-layer SNN as shown in **Figure 10**. The MNIST images are of 28x28 pixel dimensions, hence the input layer contains 784 neurons, one per image pixel. The second/hidden layer contains neurons for learning the features of the input images. The number of neurons in this layer is varied over different trials to evaluate the effectiveness of the learning rule. Finally, the third layer consists of 10

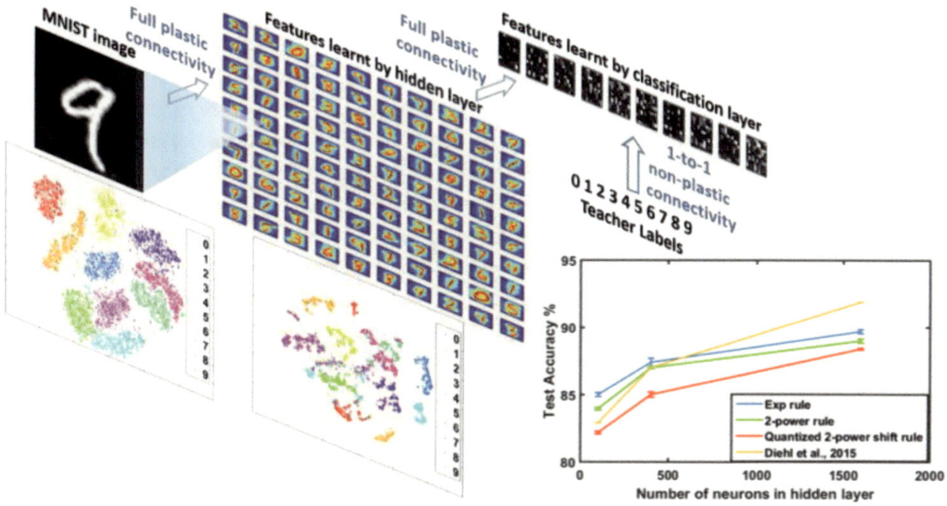

Figure 10.
MNIST SNN architecture showing connectivity, input, learnt features, labels and t-SNE visualizations, along with accuracy results [38].

neurons for classifying the input with one neuron per class. The input layer encodes the pixel intensities with varying firing rate in the range of 0 Hz – 300 Hz. Each input neuron is fully connected to the hidden layer neurons similarly each hidden layer neuron is fully connected to the output/classification layer neurons. In this network all synapses are plastic with soft WTA connectivity implemented between input layer and hidden layer neurons to facilitate different neurons to pick up shared features. On the other hand, a hard WTA connectivity exists between hidden layer and the classification layer.

A qualitative analysis of the learning rule is depicted by the t-distributed stochastic neighbor embedding (t-SNE) [61] visualizations in **Figure 10**. The t-SNE algorithm maps high dimensional data points lying on different but related low-dimensional manifolds to lower dimensions by capturing local structure present in high dimensional data. The input layer firing rate visualizations show the clustering of digit classes in 2 dimensions based on raw pixel data which has 784 dimensions. Similarly, the second visualization is made using the firing rate based on the learnt features of hidden layer as input to the t-SNE algorithm with 100 dimensions. It can be clearly seen that the STDP rule produces tight clustering of input space which is projected on to the feature space. The classification layer further groups these features to its respective classes. Networks with different number of hidden layer neurons are experimented with and the results are shown in the bottom right side of **Figure 10**. The robustness of the learning method is also demonstrated with experiments yielding similar accuracies with additive Gaussian white noise along with the use of NWTA network.

6.2 Probabilistic graphical models as SNNs

An inference network based on a probabilistic graphical model for sentence construction is created using Bayesian neurons. It consists of lexicons representing words and phrases. Here each lexicon is a WTA sub network.

The network consists of two functional sections: word sub network and phrase sub network. Each symbol neuron in word sub network represents a possible word occurrence and each symbol neuron in phrase sub network represents a possible pair of words co-occurring. The synapses between the symbol neurons represent the log conditional probabilities of words and phrases co-occurring. This network is initialized to have same intrinsic potential across all symbol neurons resulting in same initial firing rate. Based on the synaptic weights the strongly connected neurons resonate and enhance each other while laterally inhibiting other symbol neurons within the lexicon WTA network. These winning neurons proportionally excite other symbol neurons across different lexicons. In this manner the network settles on a steady state firing rate which represents a contextually correct behavior. From each lexicon of the word sub network a symbol neuron is picked with highest firing rate representing a grammatically correct semantically meaningful sentence. The WTA connections in this network perform soft WTA action there by the facilitating the retention of contextual information. **Figure 11** (a) shows the network topology. For the experiments, random documents images are picked, and fuzzy character recognition is performed. Due to the fuzzy nature, each character position will result in several possible matches hence, multiple possible matches for each word position is possible as described in [62]. An example of lexicon set is [{we, wo, fe, fo, ne, no, ns, us} {must, musk, oust, onst, ahab, bust, chat} {now, noa, non, new, how, hew, hen, heu} {find, rind, tina} {the, fac, fro, kho} {other, ether}]. The SNN after evaluating the lexicons settles on a grammatically correct sentence as [we must now find the other] as seen in **Figure 11** (b).

6.3 SNN with Backpropagation-STDP based learning

Using the learning rule presented in Section 4.4, the authors of [45] train SNNs to evaluate BP-STDP rule on the XOR problem, the iris dataset and the MNIST dataset. They show that the network can model the linearly inseparable XOR problem using an SNN with 2 input, 20 hidden and 2 output neurons. For the iris dataset they create a SNN with 4 input, 30 hidden and 3 output neurons. With this network they were able to achieve 96% accuracy which is comparable to ANN trained with traditional backpropagation with an accuracy of 96.7%. The SNN for MNIST dataset

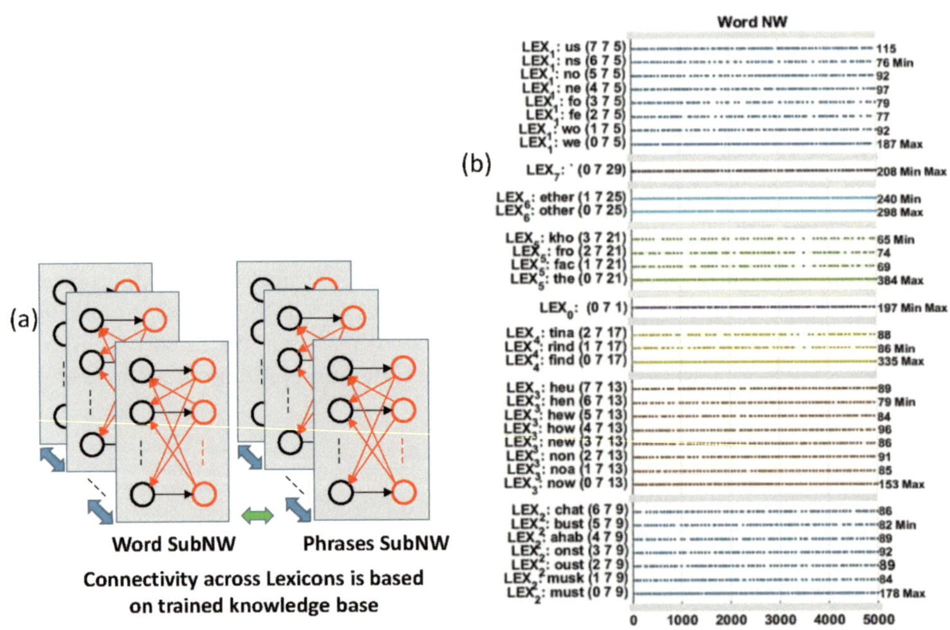

Figure 11.
(a) Sentence confabulation network, (b) confabulation results spike plot [62].

consists of 784 input neurons, 100 through 1500 hidden neurons and 10 output neurons. With this network they were able to achieve 97.2% classification accuracy.

6.4 SNNs on Neuromorphic hardware

Deep networks achieve higher accuracy in recognition tasks and in some cases outperform humans. Eedn framework is proposed in [63], which enables SNNs to be trained using backpropagation with batch normalization [64] and implement them on TrueNorth neuromorphic hardware. The Eedn trained networks are capable of achieving state-of-the-art accuracy across eight standard datasets of vision and speech. In this implementation the inference on hardware can be run at up to 2600 frames/s which is faster than real time while consuming very low power of at most 275 mW across their experiments. The network uses low precision ternary weights +1, 0 and − 1 for its synapses. A binary activation function with an approximate derivative is modeled to enable backpropagation. A hysteresis parameter is introduced in the weight update rule to avoid rapid oscillations of weights during learning. The input images are transduced by applying 12 different convolutional filter operators with binary outputs to get 12 channel input to the network as shown in **Figure 12**.

Experiments were performed on eight datasets using five different network sizes spanning across several TrueNorth chips. The results of the experiments are summarized in **Figure 13**.

Figure 12.
Example image from CIFAR10 (column 1) and the corresponding output of 12 typical transduction filters (columns 2–13) [63].

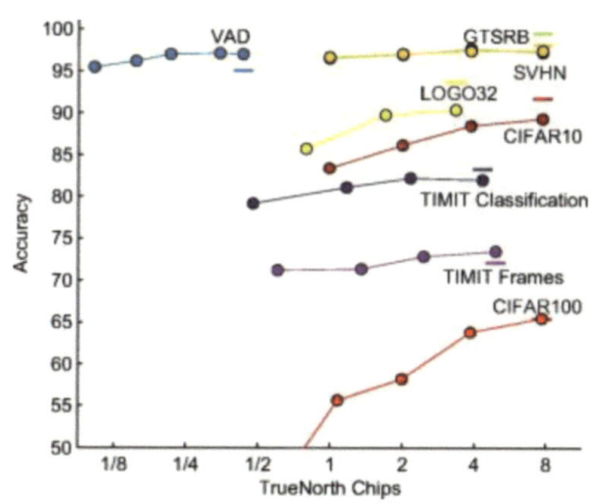

Figure 13.
Accuracy of different sized networks on eight datasets. For comparison, accuracy of state-of-the-art unconstrained approaches are shown as bold horizontal lines [63].

7. Conclusion

This chapter discussed several concepts and techniques, all of which are bio inspired. The case studies presented provide a strong basis to grasp the immense potential these algorithms provide in tackling the very complex problems of today, which were unimaginable without the advances in this field. This chapter specifically provided a beginner's guide to the field of spiking neural networks. It presented a brief overview of neuron biology and notes on popular artificial neuron models. Information representation as spikes and how to transduce real world data to spikes and vice-versa was discussed which is similar to how brain represents information. Several tools for spiking neural network modeling and evaluation were provided for wholistic understanding and for experimental evaluation of one's network models. A few case study examples are presented to understand the presented concepts and the scope of information presented in this chapter. This is an ongoing research and a very hot topic with substantially new concepts and discoveries being published every week. The motivation being the ability for machines to autonomously and efficiently perform tasks which were previously delegated to humans only along every aspect of our lives. This is a paradigm shift and research will continue to not only develop machine intelligence but also to understand the inner workings of our brains, our thoughts and advance the field of neuroscience.

Acknowledgements

This chapter represents fundamental knowledge for understanding spiking neural networks. Some of the text and images are adopted from the available research literature. Rest of the work represents authors original contributions along with the co-authors of the following research contributions [20, 38, 39, 43, 60, 62]. I am thankful for the support of Dr. Qinru Qiu from Syracuse University and her research group members specifically Amar Shrestha in contributing during the original research.

List of Abbreviations

AI	artificial intelligence
ANN	artificial neural networks
BN	Bayesian neuron
BP STDP	backpropagation-STDP
Ex	excitor neuron
LL	lower limit neuron
LTD	long-term depression
LTP	long-term potentiation
MW	moving-window
NWTA	normalized winner take all
PSTH	peri-stimulus-time histogram
SF	step-forward
SNN	spiking neural network
STDP	spike timing dependent plasticity
t-SNE	t-distributed stochastic neighbor embedding
UL	upper limit neuron
WTA	winner take all

Author details

Khadeer Ahmed
Synopsys Inc, Mountain View, USA

*Address all correspondence to: khadeer.ah@gmail.com

References

[1] Guoqing Z, Tao L. Bio-inspired autonomous navigation system for logistics mobile robots with inertial AHRS. In: 2017 IEEE 3rd Information Technology and Mechatronics Engineering Conference (ITOEC). New York City, NY, USA: IEEE; 2017. pp. 971-975

[2] Chengetanai G, O'Reilly GB. Review of swarm intelligence routing algorithms in wireless mobile ad hoc networks. In: 2015 IEEE 9th International Conference on Intelligent Systems and Control (ISCO). New York City, NY, USA: IEEE; 2015. pp. 1-7

[3] Bermejo-Busto J, Martin-Gomez C, Zuazua-Ros A, Ibanez-Puy M, Miranda-Ferreiro R, Baquero-Martin E. Improvement of a Peltier HVAC System Integrated into Building Envelopes Implementing Beehive Strategies: A Theory-Based Approach. Federacion Asociaciones Ingenieros Industriales Espana Alameda De Mazarredo. Bilbao Spain: DYNA Publishing; 2016

[4] Maass W. Networks of spiking neurons: The third generation of neural network models. Neural Networks. 1997;**10**(9):1659-1671

[5] Markram H, Gerstner W, Sjöström PJ. Spike-timing-dependent plasticity: A comprehensive overview. Frontiers in Synaptic Neuroscience. 2012;**4**:2. DOI: 10.3389/fnsyn.2012.00002. ISSN: 1663-3563. Available from: https://www.frontiersin.org/article/10.3389/fnsyn.2012.00002

[6] Kasabov NK. NeuCube: A spiking neural network architecture for mapping, learning and understanding of spatio-temporal brain data. Neural Networks. 2014;**52**:62-76

[7] iniLabs [Online]. Available: https://inilabs.com

[8] McCulloch WS, Pitts W. A logical calculus of the ideas immanent in nervous activity. Bulletin of Mathematical Biology. 1990;**52**(1–2):99-115

[9] Hodgkin AL, Huxley AF. A quantitative description of membrane current and its application to conduction and excitation in nerve. The Journal of Physiology. 1952;**117**(4):500-544

[10] Rosenblatt F. The perceptron: A probabilistic model for information storage and organization in the brain. Psychological Review. 1958;**65**(6):386

[11] Izhikevich EM. Simple model of spiking neurons. IEEE Transactions on Neural Networks. 2003;**14**(6):1569-1572

[12] Abbott LF. Lapicque's introduction of the integrate-and-fire model neuron (1907). Brain Research Bulletin. 1999;**50**(5–6):303-304

[13] Stein RB. A theoretical analysis of neuronal variability. Biophysical Journal. 1965;**5**(2):173-194

[14] Ermentrout GB, Kopell N. Parabolic bursting in an excitable system coupled with a slow oscillation. SIAM Journal on Applied Mathematics. 1986;**46**(2):233-253

[15] Fourcaud-Trocmé N, Hansel D, Van Vreeswijk C, Brunel N. How spike generation mechanisms determine the neuronal response to fluctuating inputs. The Journal of Neuroscience. 2003;**23**(37):11628-11640

[16] Jolivet R, Lewis TJ, Gerstner W. Generalized integrate-and-fire models of neuronal activity approximate spike trains of a detailed model to a high degree of accuracy. Journal of Neurophysiology. 2004;**92**(2):959-976

[17] Stevens CF, Zador AM. Novel Integrate-and-re-Like Model of Repetitive Firing in Cortical Neurons.

Rockville, MD, USA: American Physiological Society; 1998

[18] Smith GD, Cox CL, Sherman SM, Rinzel J. Fourier analysis of sinusoidally driven thalamocortical relay neurons and a minimal integrate-and-fire-or-burst model. Journal of Neurophysiology. 2000;**83**(1):588-610

[19] Izhikevich EM. Resonate-and-fire neurons. Neural Networks. 2001;**14** (6–7):883-894

[20] Ahmed K, Shrestha A, Qiu Q. Simulation of bayesian learning and inference on distributed stochastic spiking neural networks. In: 2016 International Joint Conference on Neural Networks (IJCNN). New York City, NY, USA: IEEE; 2016. pp. 1044-1051

[21] Meunier C, Segev I. Playing the Devil's advocate: Is the Hodgkin–Huxley model useful? Trends in Neurosciences. 2002;**25**(11):558-563

[22] Cassidy AS et al. Cognitive computing building block: A versatile and efficient digital neuron model for neurosynaptic cores. In: The 2013 International Joint Conference on Neural Networks (IJCNN). New York City, NY, USA: IEEE; 2013. pp. 1-10

[23] Gerstner W, Kistler WM, Naud R, Paninski L. Neuronal Dynamics: From Single Neurons to Networks and Models of Cognition. Cambridge, England, UK: Cambridge University Press; 2014

[24] Adrian ED, Zotterman Y. The impulses produced by sensory nerve-endings: Part II. The response of a single end-organ. The Journal of Physiology. 1926;**61**(2):151-171

[25] Forrest MD. The sodium-potassium pump is an information processing element in brain computation. Frontiers in Physiology. 2014;**5**:472. DOI: 10.3389/fphys.2014.00472

[26] Forrest MD. Intracellular calcium dynamics permit a Purkinje neuron model to perform toggle and gain computations upon its inputs. Frontiers in Computational Neuroscience. 2014;**8**:86. DOI: 10.3389/fncom.2014.00086

[27] Lestienne R. Determination of the precision of spike timing in the visual cortex of anaesthetised cats. Biological Cybernetics. 1996;**74**(1):55-61

[28] Mainen ZF, Sejnowski TJ. Reliability of spike timing in neocortical neurons. Science. 1995;**268**(5216):1503-1506

[29] Ponulak F, Kasinski A. Introduction to spiking neural networks: Information processing, learning and applications. Acta Neurobiologiae Experimentalis (Wars). 2011;**71**(4):409-433

[30] Stein RB, Gossen ER, Jones KE. Neuronal variability: Noise or part of the signal? Nature Reviews. Neuroscience. 2005;**6**(5):389-397

[31] Zohar O, Shamir M. A readout mechanism for latency codes. Frontiers in Computational Neuroscience. 2016; **10**:107

[32] Kim J, Kim H, Huh S, Lee J, Choi K. Deep neural networks with weighted spikes. Neurocomputing. 2018;**311**: 373-386

[33] Zeldenrust F, Wadman WJ, Englitz B. Neural coding with bursts—Current state and future perspectives. Frontiers in Computational Neuroscience. 2018;**12**:48. DOI: 10.3389/fncom.2018.00048

[34] Thorpe S, Gautrais J. Rank order coding. In: Computational neuroscience. Boston, MA, USA: Springer; 1998. pp. 113-118

[35] Cattani A, Einevoll G, Panzeri S. Phase-of-Firing Code. Ithaca, NY, USA: arXiv.org, Cornell University; 2015

[36] Montemurro MA, Rasch MJ, Murayama Y, Logothetis NK, Panzeri S. Phase-of-firing coding of natural visual stimuli in primary visual cortex. Current Biology. 2008;**18**(5):375-380

[37] Danielson NB, Zaremba JD, Kaifosh P, Bowler J, Ladow M, Losonczy A. Sublayer-specific coding dynamics during spatial navigation and learning in hippocampal area CA1. Neuron. 2016;**91**(3):652-665

[38] Shrestha A, Ahmed K, Wang Y, Qiu Q. Stable spike-timing dependent plasticity rule for multilayer unsupervised and supervised learning. In: 2017 International Joint Conference on Neural Networks (IJCNN). New York City, NY, USA: IEEE; 2017. pp. 1999-2006

[39] Shrestha A et al. A spike-based long short-term memory on a neurosynaptic processor. In: 2017 IEEE/ACM International Conference on Computer-Aided Design (ICCAD). New York City, NY, USA: IEEE; 2017. pp. 631-637

[40] Schrauwen B, Campenhout J. BSA, a fast and accurate spike train encoding scheme. In: Proceedings of the International Joint Conference on Neural Networks. Vol. 4. New York City, NY, USA: IEEE; 2003. pp. 2825-2830. DOI: 10.1109/IJCNN.2003.1224019

[41] Kasabov N et al. Design methodology and selected applications of evolving spatio-temporal data machines in the NeuCube neuromorphic framework. Neural Networks. 2016;**78**:1-14

[42] Hebb DO. The Organization of Behavior: A neuropsychological Theory. Abingdon, England, UK: Taylor & Francis; 1949

[43] Ahmed K, Shrestha A, Qiu Q, Wu Q. Probabilistic inference using stochastic spiking neural networks on a neurosynaptic processor. In: 2016 International Joint Conference on Neural Networks (IJCNN). New York City, NY, USA: IEEE; 2016. pp. 4286-4293

[44] Masquelier T, Guyonneau R, Thorpe SJ. Spike timing dependent plasticity finds the start of repeating patterns in continuous spike trains. PLoS One. 2008;**3**(1)

[45] Tavanaei A, Maida A. BP-STDP: Approximating backpropagation using spike timing dependent plasticity. Neurocomputing. 2019;**330**:39-47

[46] Haydon PG, Carmignoto G. Astrocyte control of synaptic transmission and neurovascular coupling. Physiological Reviews. 2006; **86**(3):1009-1031

[47] Xu X, Zhao Z, Li R, Zhang H. Brain-inspired Stigmergy learning. IEEE Access. 2019;7:54410-54424

[48] Stimberg M, Brette R, Goodman DFM. Brian 2, an intuitive and efficient neural simulator. eLife. 2019;**8**:e47314. DOI: 10.7554/eLife.47314

[49] Gewaltig M-O, Diesmann M. Nest (neural simulation tool). Scholarpedia. 2007;**2**(4):1430

[50] Hines ML, Carnevale NT. The NEURON simulation environment. Neural Computation. 1997;**9**(6):1179-1209

[51] Davison AP et al. PyNN: A common interface for neuronal network simulators. Frontiers in Neuroinformatics. 2009;**2**:11

[52] Furber SB et al. Overview of the spinnaker system architecture. IEEE Transactions on Computers. 2012; **62**(12):2454-2467

[53] Grübl A, Billaudelle S, Cramer B, Karasenko V, Schemmel J. Verification and Design Methods for the BrainScaleS

Neuromorphic Hardware System. arXiv Prepr. arXiv2003.11455. 2020

[54] Stöckel A. Cypress: C++ Spiking Neural Network Simulator Framework [Online]. Available from: https://github.com/hbp-unibi/cypress

[55] Vitay J, Dinkelbach HÜ, Hamker FH. ANNarchy: A code generation approach to neural simulations on parallel hardware. Frontiers in Neuroinformatics. 2015;**9**:19

[56] DeBole MV et al. TrueNorth: Accelerating from zero to 64 million neurons in 10 years. Computer (Long. Beach. Calif). 2019;**52**(5):20-29

[57] Davies M et al. Loihi: A neuromorphic manycore processor with on-chip learning. IEEE Micro. 2018; **38**(1):82-99

[58] Chou T-S et al. CARLsim 4: An open source library for large scale, biologically detailed spiking neural network simulation using heterogeneous clusters. In: 2018 International Joint Conference on Neural Networks (IJCNN). New York City, NY, USA: IEEE; 2018. pp. 1-8

[59] Catania V, Mineo A, Monteleone S, Palesi M, Patti D. Noxim: An open, extensible and cycle-accurate network on chip simulator. In: 2015 IEEE 26th International Conference on Application-Specific Systems, Architectures and Processors (ASAP). New York City, NY, USA: IEEE; 2015. pp. 162-163

[60] Ahmed K, Shrestha A, Wang Y, Qiu Q. System design for in-hardware stdp learning and spiking based probablistic inference. In: 2016 IEEE Computer Society Annual Symposium on VLSI (ISVLSI). New York City, NY, USA: IEEE; 2016. pp. 272-277

[61] van der Maaten L, Hinton G. Visualizing data using t-SNE. Journal of Machine Learning Research. 2008;**9** (Nov):2579-2605

[62] Qiu Q, Li Z, Ahmed K, Li HH, Hu M. Neuromorphic acceleration for context aware text image recognition. In: 2014 IEEE Workshop on Signal Processing Systems (SiPS). New York City, NY, USA: IEEE; 2014. pp. 1-6

[63] Esser S, et al. Convolutional Networks for Fast, Energy-Efficient Neuromorphic Computing. arXiv. arXiv Prepr. arXiv1603.08270. 2016

[64] Ioffe S, Szegedy C. Batch Normalization: Accelerating Deep Network Training by Reducing Internal Covariate Shift. arXiv Prepr. arXiv1502.03167. 2015

Chapter 6

Unsteady Aerodynamics of Highly Maneuvering Flyers

Mohamed Yehia Zakaria

Abstract

In this chapter, a set of analytical aerodynamic models, based on potential flow, that can be used to predict the unsteady lift response during pitching maneuvers are presented and assessed. The result examines the unsteady lift coefficients experienced by a flat plate in high-amplitude pitch ramp motion. The pitch ramps are chosen based on two ramp pitch maneuvers of a maximum amplitudes of 25 and 45 degrees starting from zero degree. The aim is investigate the use of such classical models in predicting the lift dynamics compared to a full physical-based model. Among all classical methods used, the unsteady vortex lattice method (without considering the leading edge vortex) is found to be a very good predictor of the motion lift dynamic response for the 25° ramp angle case. However, at high pitch maneuvers (i.e.,the 45° ramp angle case), could preserve the response pattern with attenuated amplitudes without high computational burden. These mathematical analytical models presented in this chapter can be used to obtain a fast estimate for aircraft unsteady lift during pitch maneuvers instead of high fidelity models, especially in the early design phases.

Keywords: canonical maneuvers, pitching maneuvers, unsteady aerodynamics, unsteady lift response

1. Introduction

Loops, barrel rolls and pitch maneuvers are impressive aerial stunts. But even during the most intense in-air aerobatics, most planes are still constrained by aerodynamics. The air flowing over their wings gives them the lift to stay aloft and they control their movement by altering the surfaces that air flows over. The quick the rate of movement for the control surface, a fast response from the aircraft to change attitude. Pilots can pull off moves with precise control in conditions that would leave other aircraft hopelessly plummeting towards the ground. For fighter aircraft, there are numerous maneuvers can be done by the pilot to increase the aircraft maneuverability. These maneuvers such as, Cobra, Mango flip, high pass alpha that can save pilot's life during a dog fight (see **Figure 1**). Nowadays, unmanned aeriel vehicles autopilots can perform these maneuvers to an extent. Consequently, in order assure that UAVs could perform such maneuvers, one may need to relax the quasi-steady modeling to an unsteady nonlinear model to deal with these abrupt changes in attitude. Prediction of dynamic lift response of Harsh maneuvers for flying vehicles necessitate a compact aerodynamic modeling. For instance, pitching

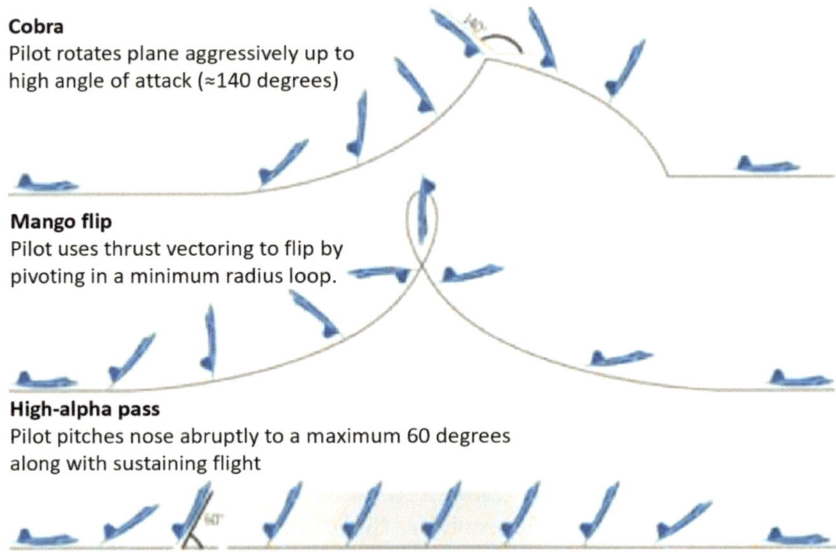

Figure 1.
High alpha Fighter's aircraft maneuvers.

maneuvers for fighter aircrafts (ex. F35 - SU-57) with specified handling qualities stimulate the idea to impose new modeling techniques to be applied on UAVs. The unsteady lift response plays an important role to control the vehicle at such low speeds. Escaping from a flying threat, first performed by Soviet test pilot Viktor Pugachoyov in 1989, the maneuver that would go on to be called "Pugachev's Cobra" is one of the building blocks that makes up many other more complicated supermaneuvers. During flight, the pilot pulls back to an absurd angle of attack, taking the nose of the aircraft completely vertical or even beyond. From here, one of two things can happen. In a plane without thrust vectoring but with a thrust-to-weight ratio higher than one, the drag towards the tail of the plane can be used to pitch the nose forward again. If the plane does have thrust vectoring, that can help the re-orientation even more. But either way, the engines are firing hard enough the entire time to maintain the jet's altitude despite the loss of speed and lift.

After few years, a German test pilot Karl-Heinz Lang performed the Herbst Maneuver in 1993. The Herbst Maneuver is basically Pugachev's Cobra with a bit of a twist. Instead of just pulling up and going forward again, the Herbst Maneuver has the pilot roll the plane (experimental X-31) a bit while its nose is pointed at the sky, so that when the nose comes back down, the plane is pointed in a different direction. On the other hand, such maneuvers are also possessed by birds and flapping insects. They can twist their wings at high angles of attack while flapping their wings without approaching stall. This is known as non-conventional lifting mechanisms invoked from biomemetics in order to perform such maneuver with a stabilized flight (i.e. vibrational stabilization). In preliminary design of UAVs, potential flow models are used as a start point to ensure acceptable estimates for aerodynamic forces and moments. A recent motivation is devoted towards designing flight control systems that can achieve harsh maneuvers such as perching and sudden landing for fixed wing MAV's [1, 2]. Bird perching is considered one of the most fascinating landing and decelerating maneuvers. **Figure 2(a)** shows a tailed swallow feeding a chick by pitching its wing at high angle of attack. For specific missions, such maneuver is useful for both flapping-wing and fixed-wing MAVs.

For classical unsteady aerodynamic models, Theodorsen [5], Wagner [6] and others have been studied extensively the classical theories of unsteady aerodynamics to be employed in the aeroelasticity field. However, aerodynamic models of harsh maneuvers characterized by sharp pitch rates and amplitudes still present a challenge in modeling. While advances in computational fluid dynamics and experimental methods have opened the study of these maneuvers as such a low-fidelity analytical modeling for rigorous prediction is still forthcoming. Roderich et al. [4] performed experiments for touchdown to take-off for a very basic glider as shown in **Figure 2(b)**.

In the last two decades, there have been several efforts exerted on unsteady aerodynamic modeling based on potential flow theories as well as modified thin airfoil theory to simulate the wing motion for an arbitrary input [7, 8]. The AIAA Fluid Dynamics Technical Committee's (FDTC) Low Reynolds Number Discussion Group introduced some cases for the assessment of experimental efforts [9], on large amplitude pitching maneuvers. The proposed motions are used as a benchmark for obtaining analytical and phenomenological models, in which a ramp up, hold, and ramp-down motions are analyzed using theory and numerical computations [10–14] Theodorsen's and Wagner's Inviscid theories are purely proper only for small amplitude oscillations associated with planar wakes. However, a tremendous work has shown that these methods remain substantially accurate even at moderate amplitudes and high frequencies. The results obtained by Ramesh et al. [9] during the hold and downstroke show that the aerodynamic forces are dominated by a deep-stall as well as leading edge vortex (LEV). The shedding effects were seen from the vorticity and dye injection plots from his experimental results. These results proved that viscous state indicate that the inviscid assumptions are insufficient for modeling the hold and downstroke portions of the motion and adequate for capturing the lift time history during the ramp phase.

A tremendous work was done based on nonlinear unsteady reduced order modeling to solve flow at high frequencies [8, 15–19]. The recent work done by Yuelong et al. [20] examined the unsteady forces and moment coefficients obtained by a thin airfoil in a pitch ramp high-amplitude motion. Wind tunnel experiments have been conducted at Reynolds number ($Re = 45 \times 10^4$), using a rigid flat-plate

(a) (b)

Figure 2.
Example of bird perching and successful experiments based on perching manoeuver. (a) A wire-tailed swallow feeding a re-cently edged chick [3] (a) A wire-tailed swallow feeding a re-cently edged chick [3]. (b) A basic glider, manually thrown and con-trolled by perching [4].

model. Forces have been measured for reduced pitch rates ranging from 0.01 to 0.18 reduced frequency ($k = \omega c/2U_\infty$) along with four maximum pitch angles (30°; 45°; 60°; 90°) at different pivot axis locations. The results show that the unsteady aerodynamics is limited to a delayed stall effect for reduced pitch rates lower than k = 0.03. At higher pitch rates, the unsteady aerodynamic response is associated with a formation of circulation, which in turn increases with the pitch rate and the distance between the pivot axis and the 3/4-chord location. An enhanced response was noted in the normal force and moment coefficients due to these circulatory effects. These overshot is slightly reduced for a flat plate with a finite aspect ratio near eight compared to two-dimensional configuration. The authors proposed a new time-dependent model for both lift and moment coefficients. The model based on the Wagner function and a time-varying input along with nonlinear variation of the quasi steady aerodynamics. A satisfactory results for 0° to 90° pitch ramp motions were compared with experiments for different pivot locations and various circulation intensity based on pitch rates.

On the other hand, fluid structure interaction modeling became essential for solving flow around vibrating and rotating structure [8, 21–23]. Modeling such moving bodies requires aerodynamic unsteady nonlinear models to assure accuracy in modeling results rather than using quazi-steady models. Carlos et al. [24] work discuses modeling and analyzing procedures of the non-linearities induced by the flow-structure interaction of an energy harvester consisting of a laminated beam integrated with a piezoelectric sensor. The cantilevered beam and the piezoelectric lamina are modeled using a nonlinear finite element approach, while unsteady aerodynamic effects are described by a state-space model that allows for arbitrary nonlinear lift characteristics.

The major contribution about the classical unsteady formulations discussed in the literature is the inefficacy to account for a non-conventional lift curve, such as LEV effects and dynamic stall contributions. Taha et al. [7] developed a state space model that captures the nonlinear contributions of the LEV in an unsteady fashion. However, their underpinning dynamics is linear: convolution with Wagner's step response. Consequently, there is a considerable gap in the literature for consolidating low fidelity models for predicting accurate lift forces associated with these large-amplitude maneuvers. An analytical unsteady nonlinear aerodynamic model that can be used to characterize the local and global nonlinear dynamic characteristics of the airflow is a mandatory task for aerodynamicists. Developing such a model will be indispensable for multidisciplinary applications (e.g., dynamics, control and aeroelasticity).

The chapter investigates and assesses relevant classical analytical models in solving lift response for pitching maneuovers. In doing so, Theodorsen, Wagner and Unsteady vortex lattice methods are used to predict the lift dynamics, then the results are compared with the experimental data presented by Ramesh et al. [9]. Also, the work proposed a simple time-dependent model in order to predict the lift response for a two dimensional wing performing rapid pitch motion. In addition, the results provide a comparison with numerical simulation using the unsteady vortex lattice method. The aerodynamic system receives the time histories of angle of attack, quasi-steady lift as inputs and produces the corresponding total unsteady lift as output. In the following sections, each presented model will be explained in detail. The chapter is organized as follows. The adopted motion kinematics are presented in Section 2. Aerodynamic classical models are reported in Section 3, along with the effect of reduced pitch rate and pivot axis location. In Section 4, the effect of pitch amplitudes on the unsteady lift coefficient is investigated by comparing the obtained results using two different pitch amplitudes with the experimental results [9].

2. Motion kinematics

In order to explore the non-periodic motions of wings rapid manouevers, the ramp-hold-return motions were proposed by the AIAA FDTC Low Reynolds Number Discussion Group [25]. The smoothed ramp motion proposed by Eldredge's canonical formulation [12] is used in this work as a reference case for comparison. Here, the experimental work done by Ramesh et al. [13] is considered as a benchmark. Variations of this motion are considered by varying the pitch amplitude (25° and 45°) at a Reynolds number of 10,000. **Figures 2** and **3** show a schematic of the pitch motion variables and the two studied maneuvers versus the non-dimensional time, respectively. **Figure 4** shows the ramped motion for a maximum amplitude of 25° versus the corresponding effective angle of attack and the local angle of attack at the 3/4 chord location as suggested by Pistolesi theorem [26].

To avoid any numerical instabilities, (e.g., dirac-delta function spikes in the calculation of the added mass force) all motions are smoothed based on a smoothing parameter introduced by Elderedge [12]. For a ramp going from 0 degrees angle of attack to 25 or 45 degrees, the first 10% (2.5 or 4.5 degrees) can be replaced with a sinusoidal tangent to the baseline ramp, and similarly in approaching the "hold" portion at the maximum amplitude angle of attack, consequently again on the downstroke. This treatment avoids a piece-wise linear fit which has discontinuities in the angle derivatives. The smoothing function G(t) is defined as:

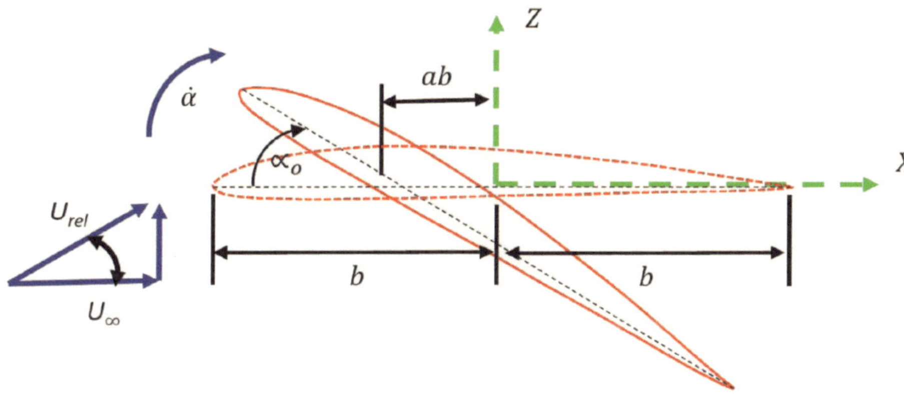

Figure 3.
Pitching motion nomenclature and motion variables (a = 1 is the leading edge pivot, a = 0 is the mid chord pivot and a = −1 is trailing edge pivot).

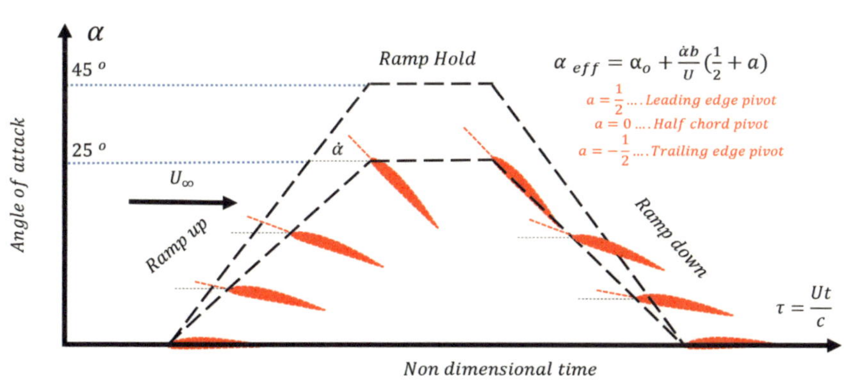

Figure 4.
The proposed ramp maneuver with a maximum amplitudes of 25° and 45° and pitch rates of 0.2 and 0.4, respectively.

$$G(t) = ln\left(\frac{cosh\left(aU_\infty(t-t_1)/c\right) cosh\left(aU_\infty(t-t_4)/c\right)}{cosh\left(aU_\infty(t-t_1)/c\right) cosh\left(aU_\infty(t-t_4)/c\right)}\right) \tag{1}$$

where a is the smoothing parameter and is taken to be 11, t_1 through t_4 are the transition times and a pitch amplitude angle A. As such, the smoothed angle of attack can be written as:

$$\alpha(t) = A\frac{G(t)}{max\,(G(t))} \tag{2}$$

3. Classical models

In order to analytically describe the generated lift force due to pitching maneuvers, a well established models were introduced. In this section, a detailed description of these models is discussed and explained in a straight forward manner.

3.1 Theodorsen model

The tremendous work done by Wagner [6], Prandtl [27], Theodorsen [28] and Garrick [29] described some fundamental physical concepts in understanding and modeling the unsteady aerodynamics. These concepts are usually incorporated with a potential flow approach and small disturbance theory to obtain analytical expressions of flow quantities. The unsteady lift on a harmonically oscillating airfoil in incompressible flow has been studied by Kussner and Schwarz [30], but the most well known solution is due to Theodorsen [5]. The lift on a thin rigid airfoil undergoing oscillatory motion can be written as:

$$L = \underbrace{\pi\rho b^2\left(\ddot{h} + U_\infty\dot{\alpha} + ba\ddot{\alpha}\right)}_{Added\quad mass} + \underbrace{2\pi\rho U_\infty b\left(\dot{h} + U_\infty\alpha + b\left(\frac{1}{2}+a\right)\dot{\alpha}\right)}_{Quasi\quad steady}C(k) \tag{3}$$

or in normalized form,

$$C_L = \frac{\pi b}{U_\infty^2}\left(\ddot{h} + U_\infty\dot{\alpha} + ba\ddot{\alpha}\right) + 2\pi C(k)\left(\frac{\dot{h}}{U_\infty} + \alpha + b\left(\frac{1}{2}+a\frac{\dot{\alpha}}{U_\infty}\right)\right) \tag{4}$$

where, $\ddot{h}$ and $\ddot{\alpha}$ are plunging and pitching accelerations respectively. The first group of terms are the noncirculatory components which account for the inertia of fluid (added mass force). The second group of terms are the circulatory components, where C(k) accounts for the influence of the shed wake vorticity (lift deficiency factor). Since Theodorsen function necessitates a periodic motion for its input parameters (e.g. angle of attack or quasi steady lift), a Fourier transform should be applied to the pitch ramp maneuver under study. The effective angle of attack of the proposed ramp pitch motion can be written as:

$$\alpha_{eff} = \alpha + \dot{\alpha}\left(\frac{1}{2}+a\right)\frac{b}{U_\infty} \tag{5}$$

Two approaches were undertaken to test the transformed input functions for Theodorsen classical unsteady model as follows:

- Fourier series approach

By applying Fourier series for the given effective maneuver angle of attack and considering Theodorsen function C(k) such that:

$$A_{c(k)} = |C(k)|, and\ \phi = \angle(C(k)) \quad (6)$$

where $A_{c(k)}$ is the absolute value (amplitude) and ϕ is the phase angle. The circulatory lift given after applying Fourier series is given by:

$$L_{Circ} = \pi\rho U^2 c a_o + \rho U^2 c\left(A_{C(k)}\right)\left(a_n \cos\left(\omega t + \phi\right) + b_n \sin\left(\omega t + \phi\right)\right) \quad (7)$$

The non-circulatory lift part [31] is given by:

$$L_{Non-Circ} = -\pi\rho b^2\left[U\dot{\alpha} + \ddot{\alpha}ab\right] \quad (8)$$

- Fast Fourier transform

The Fast Fourier Transform of the effective angle of attack is written as:

$$\alpha_{eff}(\omega) = \int_0^\infty \alpha_{eff}(t)e^{\ i\omega t}dt \quad (9)$$

and the circulatory component of lift based on FFT is given by:

$$L_{Circ-FFT} = \frac{1}{2}\rho U^2 c\int_{-\infty}^{\infty} \alpha_{eff}(w)C(k)e^{i\omega t}d\omega \quad (10)$$

It should be noted that practically, this Fourier transform approach will be implemented numerically using discrete fourier transform. However, discrete Fourier transform in contrast with the exact Fourier transform (Fourier integral) will necessarily ignore some frequency contents due to the integration limits between $-\infty$ to $+\infty$.

3.2 Wagner step response and Duhamel superposition principle

Using Wagner's linear step response, the Duhamel principle can be used to include the unsteady effects in an exact form such as a finite-state aerodynamic models suitable for aeroelastic problems and flight mechanics simulations. Wagner [6] obtained the time dependant-response of the lift on a flat plate due to a step input (indicial response problem). Garrick [29] showed that by using Fourier transformation, Wagner function, $W(s)$, and Theodorsen function, $C(k)$ can be related together. Wagner [6] determined the circulatory lift due to a step change in the wing motion. The unsteady lift is then written in terms of the static lift as:

$$\ell(s) = \ell_s W(s) \quad (11)$$

where the non-dimensional time S is defined as $S = \frac{2U_\infty t}{c}$ for constant free-stream velocity U_∞.

By knowing the indicial response for a linear dynamical system, the response due to arbitrary motion (input) can be described as an integral (superposition) using the indicial response and an input varies with time. The variation of the circulatory lift for an arbitrary change in the angle of attack is given by:

$$\ell(s) = \pi\rho U^2 c\left(\alpha(0)W(s) + \int_0^s \frac{d\alpha(\sigma)}{d\sigma} W(s-\sigma)d\sigma\right) \tag{12}$$

We note that $W(s)$ can also be used as an indicial response to aerodynamic inputs other than the angle of attack. Van der Wall and Leishman [32] used it as an indicial response to the wing normal velocity, $w = U\alpha$, in the case of time-varying free stream. For a relatively high angle of attack, the Duhamel superposition is performed using a more exact normal velocity $w = U\sin\alpha$. Eq. (11) is then re-written as

$$\ell(s) = \pi\rho U(s)c\left(U(0)\sin\alpha(0)W(s) + \int_0^s \frac{d(U(\sigma)\sin\alpha(\sigma))}{d\sigma} W(s-\sigma)d\sigma\right) \tag{13}$$

This equation is usually used in dynamic stall models where relatively high angles of attack are encountered, e.g., the Beddoes-Leishman dynamic stall model [33].

3.3 State space finite model

RT Jones proposed an approximate expression for Wagner function as follows:

$$\phi(s) = 1 - A_1 e^{-c_1 s} - A_2 e^{-c_2 s} \tag{14}$$

where $A_1 = 0.165, A_2 = 0.335, c_1 = 0.0455, c_2 = 0.3$ and s is the reduced time parameter and is given by $U_\infty t/b$. By taking the Laplace transform with an operator P:

$$\phi(P) = \frac{1}{P} - \frac{A_1}{P + \frac{c_1 U_\infty}{b}} - \frac{A_2}{P + \frac{c_3 U_\infty}{b}} \tag{15}$$

the transfer function is then written as:

$$G(P) = \frac{Y(P)}{\alpha_{eff}(P)} = \frac{\phi(P)}{1/P} = 1 - \frac{A_1\ P}{P + \frac{c_1 U_\infty}{b}} - \frac{A_2 P}{P + \frac{c_2 U_\infty}{b}} \tag{16}$$

$$G(P) = \frac{\left(P + \frac{c_1 U_\infty}{b}\right)\left(P + \frac{c_2 U_\infty}{b}\right) - A_1 P\left(P + \frac{c_2 U_\infty}{b}\right) - A_2 P\left(P + \frac{c_1 U_\infty}{b}\right)}{\left(P + \frac{c_1 U_\infty}{b}\right)\left(P + \frac{c_2 U_\infty}{b}\right)} \tag{17}$$

$$G(P) = \frac{(1 - A_1 - A_2)P^2 + \left(\frac{c_1 U}{b}(1 - A_2) + \frac{c_2 U}{b}(1 - A_1)\right)P + \frac{c_1 c_2 U^2}{b^2}}{P^2 + (c_1 + c_2)\frac{UP}{b} + \frac{c_1 c_2 U^2}{b^2}} \tag{18}$$

To determine a second-order state-space realization of the transfer function in Eq. 17 can be written as:

$$\frac{Y}{\alpha_{eff}} = \frac{b_2 P^2 + b_1 P + b_0}{P^2 + a_1 P + a_0} = \frac{Y}{X}\frac{X}{\alpha_{eff}} \tag{19}$$

where X is the internal states of the system, which is related to the input via these coefficients $a_o = \frac{(C_1+C_2)U_\infty}{b}$, $a_1 = \frac{(C_1 C_2)U_\infty^2}{b^2}$, $b_o = \frac{(C_1 C_2)U_\infty^2}{b^2}$, $b_1 = \left(\frac{C_1 U_\infty}{b} + \frac{C_2 U_\infty}{b} - \frac{A_1 C_2 U_\infty}{b} - \frac{A_2 C_1 U_\infty}{b}\right)$, $b_2 = 1 - A_1 - A_2$ as follows:

$$\frac{X}{\alpha_{eff}} = \frac{1}{P^2 + a_1 P + a_0} \tag{20}$$

$$\frac{Y}{X} = \frac{b_2 P^2 + b_1 P + b_0}{1} \tag{21}$$

and to the output via:

$$XP^2 + Xa_1 P + Xa_0 = \alpha_{eff} \tag{22}$$

then applying Laplace inverse we get:

$$\ddot{X} + a_1 \dot{X} + a_o X = \alpha_{eff} \tag{23}$$

then let $X_1 = X$ and $X_2 = \dot{X}$.
Also,

$$Y = Xb_2 P^2 + Xb_1 P + b_0 X = b_2 \ddot{X} + b_1 \dot{X} + b_o X \tag{24}$$

Hence,

$$Y = b_2 \left(\alpha_{eff} - a_o X_1 - a_1 X_2\right) + b_1 X_2 + b_o X_1 \tag{25}$$

By writing these equation in a matrix form, we obtain

$$\frac{d}{dt}\begin{pmatrix} X_1 \\ X_2 \end{pmatrix} = \begin{bmatrix} 0 & 1 \\ -a_o & -a_1 \end{bmatrix}\begin{pmatrix} X_1 \\ X_2 \end{pmatrix} + \begin{pmatrix} 0 \\ 1 \end{pmatrix}\alpha_{eff} \tag{26}$$

$$y = [b_o - b_2 a_o \; b_1 - b_2 a_1]\begin{pmatrix} X_1 \\ X_2 \end{pmatrix} + (b_2)\alpha_{eff} \tag{27}$$

then by applying the quasi-steady lift expression, we have;

$$L_{QS} = \rho U_\infty \Gamma = 2\pi \rho U_\infty b \, W_{3/4} \tag{28}$$

where $W_{3/4}$ is the normal velocity component and is given by:

$$\begin{aligned} W_{3/4} &= U_\infty \sin(\alpha) + \dot{\alpha}\left[\frac{b}{2} + a\right] \\ L_c(t) &= 2\pi \rho U_\infty b [b_o - b_2 a_o \; b_1 - b_2 a_1]\begin{pmatrix} X_1 \\ X_2 \end{pmatrix} + [b_2] W_{3/4} \end{aligned} \tag{29}$$

3.4 Unsteady vortex lattice method (UVLM)

The unsteady Vortex lattice methods (UVLMs) are well suited to the bio-inspired flight problems because they can account for the circulation distribution variations on wings, the velocity potential time-dependency, and the shedding of wake downstream. Although they are considered low fidelity models, they may be extended to capture unconventional lift mechanisms such as leading edge vortex [34–36]. These discrete vortex models are widely used in modeling aerodynamics of aircraft and rotorcraft analysis, compared to computational fluid dynamics (CFD) models which are more computationally expensive [37]. The use of UVLM method is now a powerfull tool in hand for aerodynamicists for its ease implementation even for complex shapes.

Zakaria et al. [8] used UVLM to model the aerodynamic loading on different Samara leaves (Maple seeds) during their steady state flight. The results were verified with experiments. Parameters including the drop speed, angular velocity and coning angle for different sets of Maple Samaras were determined from experiments. The aerodynamic loads were calculated using UVLM against the forces required for maintaining a steady state flight as obtained from the experiment. Consequently, the UVLM approach yields adequate aerodynamic modeling features that can be used for more accurate flight stability analysis of the Samara flight or of decelerator devices inspired by such flight. Also, Simon et al. [38] showed that by imposing an arbitrary input as a control surface deflection to an unsteady VLM suitable for efficient aerodynamic loads analysis within aeroelastic modeling, analysis and optimization frameworks for preliminary aircraft design. By using a continuous time state space aerodynamic model is extended for accepting arbitrary motion, control surface deflection and gust velocities as inputs. Their results showed good agreement for a large range of reduced frequencies. Accepting arbitrary motion, control surface deflection and gust velocities as inputs.

The (UVLM) divides the lifting surface into panels. A point vortex is then associated with each of these panels. The center of this ring is set at the 1/4 of the panel chord length. One collocation point is set in each panel at the 3/4 of the panel length, and the panel normal vector is calculated in this point as shown in **Figure 5**.

The UVLM model is based on the following assumptions:

- No penetration boundary condition.
- Kelvin Circulation Theorem (Conservation of Circulation).
- Vortices is convected by local velocities. (Wake deformation)

The velocity induced by all the vortex points, including the shed vorticies through the wake, is calculated at each control point and the no-penetration kinematic boundary condition is applied to calculate vortex intensity on each panel. At each time step, there are (m + 1) unknowns (m $\gamma_{boundvortices}$'s and $\gamma_{atrailingedgevortex}$), then (m + 1) equations are needed for closure. For the no-penetration boundary condition at m control points, we have:

$$V_{cp}^{(n)} = V_{air}^{(n)}\Big|_{cp} \tag{30}$$

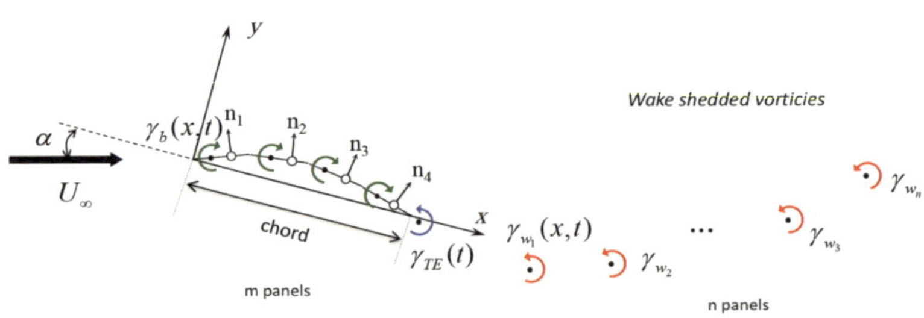

Figure 5.
A schematic diagram showing the panels on the airfoil camber and the shedded vortices used in UVLM modeling.

From Kelvin's circulation theorem, we have:

$$\sum \Gamma_{tot} = \sum \gamma_b + \gamma_w = 0 \Rightarrow \sum \gamma_b + \gamma_{TE} = -\Gamma_w \tag{31}$$

where Γ_w is the sum of all wake vortices, which are know from previous time steps. As such, we obtain the following linear system,

$$\begin{pmatrix} a_{1,1} & a_{1,2} & \dots & a_{1,MN} \\ a_{2,1} & a_{2,2} & \dots & a_{2,MN} \\ \vdots & \vdots & \ddots & \vdots \\ a_{MN,1} & a_{MN,2} & \dots & a_{MN,MN} \end{pmatrix} \begin{Bmatrix} \Gamma_1 \\ \Gamma_2 \\ \vdots \\ \Gamma_{MN} \end{Bmatrix} = \begin{Bmatrix} RHS_1 \\ RHS_2 \\ \vdots \\ RHS_{MN} \end{Bmatrix}, \tag{32}$$

where a_{K_1,K_2} are the influence coefficients from the point vortex K_2 at the control point K_1 and it is equal to the normal velocity that the point vortex induces at the control point $\Gamma_{K_2} = 1$. Each element on the right hand side is $RHS_K = -\left(\vec{v} + \vec{v}_w\right)_K \cdot \vec{n}_K$, where $\vec{v}$ is the wind flow velocity relative to the surface and $\vec{v}_w$ is the velocity induced at the control point K by all the other vortex point in the wake created before the time t. In order to satisfy the unsteady Kutta condition, the wake is created at each instant of time at the trailing edge by shedding a new vortex that has an intensity equal to the bound vortex on the panel along the trailing edge. At each instant of time all the points in the wake generated in previous steps are convected downstream following the induced velocity generated by all the vortices on the surface and through the wake. The velocity induced by each vortex is computed by using the Biot Savart law. This induced velocity is inversely proportional to the distance between the vortex location and the control point where the velocity is calculated. Having solved the linear system [39] in the bound vorticity, the pressure difference through the bound vortex sheet is computed based on the unsteady Bernoulli's equation. More details can be found in [40].

The unsteady aerodynamic loads can be calculated from the circulation Γ_K of the Kth panel and its time rate of change [40]. Using the unsteady Bernoulli equation,

$$\frac{\Delta P(x)}{\frac{1}{2}\rho U_\infty^2} = \rho\left(\frac{d}{dt}\int_0^c \gamma_b(x)dx + U_\infty \gamma_b(x)\right) \tag{33}$$

the unsteady pressure difference on the Kth panel is given by,

$$\left(\frac{p_l - p_u}{\rho}\right)_K = \left(\frac{\Delta p}{\rho}\right)_K = \left(\frac{V_u^2 - V_l^2}{2}\right)_K + \left(\frac{\partial \phi_u}{\partial t}\right)_K - \left(\frac{\partial \phi_l}{\partial t}\right)_K, \tag{34}$$

where p denotes the static pressure, V is the tangent velocity, ϕ is the velocity potential, and the subscripts u and l are used to represent the upper and lower surfaces, respectively.

From the definition of circulation, we have:

$$\left(\frac{\partial \phi_u}{\partial t}\right)_K - \left(\frac{\partial \phi_l}{\partial t}\right)_K = \frac{\partial \Gamma_{i,j}}{\partial t} = \frac{\Gamma_{i,j}(t) - \Gamma_{i,j}(t-1)}{\Delta t}, \tag{35}$$

for i = 1, 2, ..., M, and by applying the Kutta-Joukowski theorem, the normal force on each panel is obtained from:

$$\vec{F}_{N_K} = -(\Delta p \Delta S)_{i,j} \vec{n}_{i,j} \quad , \tag{36}$$

where ΔS is the area of each panel.

3.5 Models comparison

In order to summarize the merit of the proposed classical potential models for solving high pitch maneuvers, **Table 1** is shown. **Table 1** represents the key parameters for each model in the sense of input motion, nonlinearity, wake deformation and camber variation for flying vehicles. The merit of each model is how one can apply simple analytical equation to solve such maneuver.

4. Maneuver case studies results

4.1 Case 1: Pitch ramp $\alpha_o = 25^o$

4.1.1 Leading edge pivot

Figures 6–11 show a comparison between the proposed models discussed above for different ramp amplitudes and hinge locations. A physical interpretation for the jump and attenuated lift peaks show four flow events as reported by Ramesh et al. [9] as follows: (i) onset of flow separation at the ramp start ($\tau = 1$), (ii) the formation of a leading edge vortex ($\tau = 1 - 3$), (iii) ramp hold ($\tau = 3 - 4$) and (iv) detachment of the leading-edge vortex ($\tau = 4 - 6$).

Figure 6 shows the ramp pitch motion with an amplitude of 25° about a leading edge hinge. Almost all the theoretical models have the same jump during the transition of each event start and end positions during the whole ramp manoeuver compared to the experimental results. During upstroke, $\tau = 1 - 3$, a very good match is found between the experimental results and the UVLM results. The prediction of the quasi-steady model is higher compared to the experimental results which is expected as it lacks the dynamics of the flow and is based only the static behavior of the generated lift. On the other hand, all other presented models show an attenuated lift response during the ramp-up phase. During the ramp hold period, $\tau = 3 - 4$, a very good agreement between all the models and the experiments except the quasi-steady and Theodorsen FFT based model. During the ramp-down phase, the UVLM model matches very well with the experimental results preserving the lift dynamics. On the other hand, all models over predict the lift coefficient except the quazi-steady model shows a lower lift coefficient. As reported by Yu et al. [14], the reason for this discrepancy could be attributed to the sensitivity of these models to capture the LEV de-attachment and the lift decrease during this phase. Based on the observations of Ramesh et al. [9], this behavior points to an

Models	Input motion	Nonlinearity	wake deformation	Camber variation
Theodorsen	Harmonic	Geometric	Flat	×
Wagner	Step input	Geometric	×	×
State space	Arbitrary	×	×	×
UVLM	Arbitrary	√	√	√

Table 1.
Classical aerodynamics proposed models for solving pitching maneuvers.

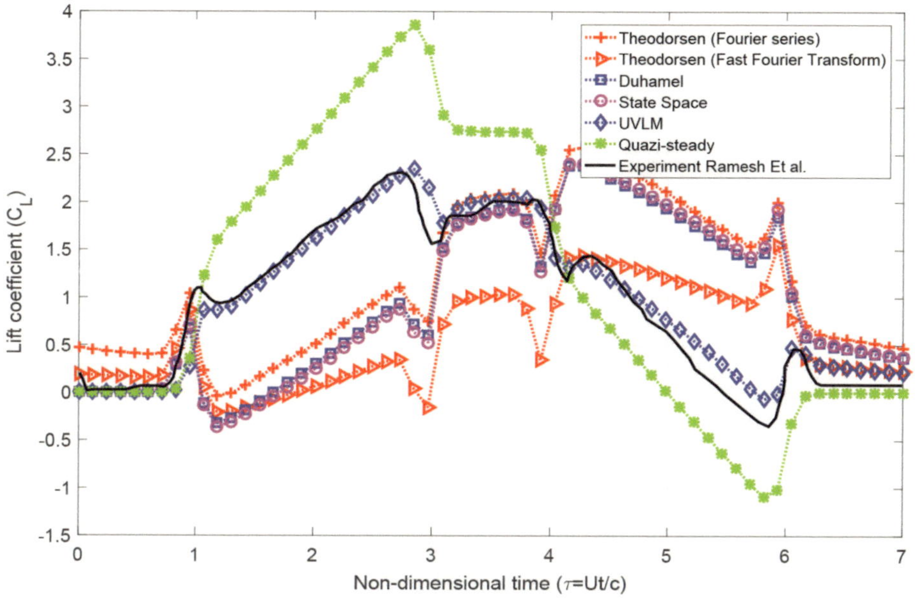

Figure 6.
Comparison for the proposed models and experimental work done by Ramesh et al. with ramp rate of 0.2 and amplitude 25° at the leading edge hinge location.

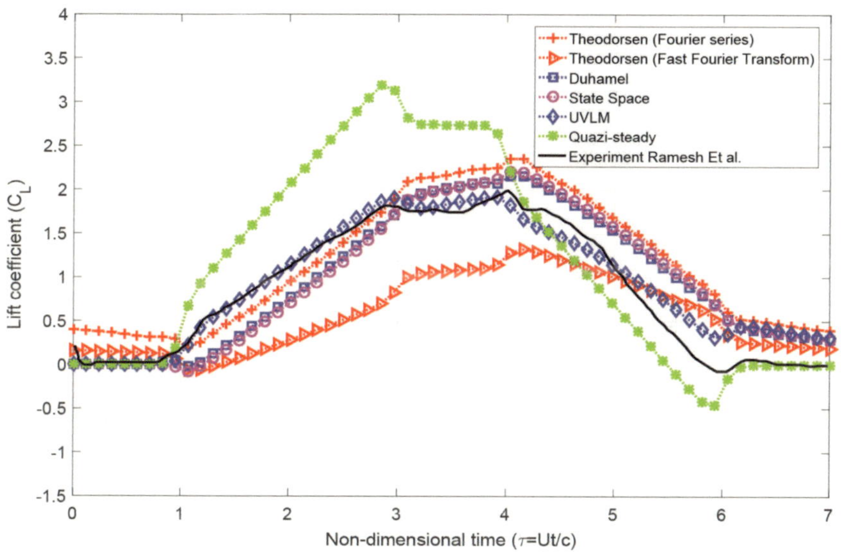

Figure 7.
Comparison for the proposed models and experimental work done by Ramesh et al. with ramp rate of 0.2 and amplitude 25° at the half chord hinge location.

important aspect, where the lift dynamics results in a considerable delay; i.e. the lift response does not depend on the past history (memory effects).

4.1.2 Half chord pivot

Figure 7 shows the ramp with amplitude of 25° at the mid chord hinge location. The results show a good agreement with the experimental results by having the same lift response slope except for Thoedorsen model based on FFT model

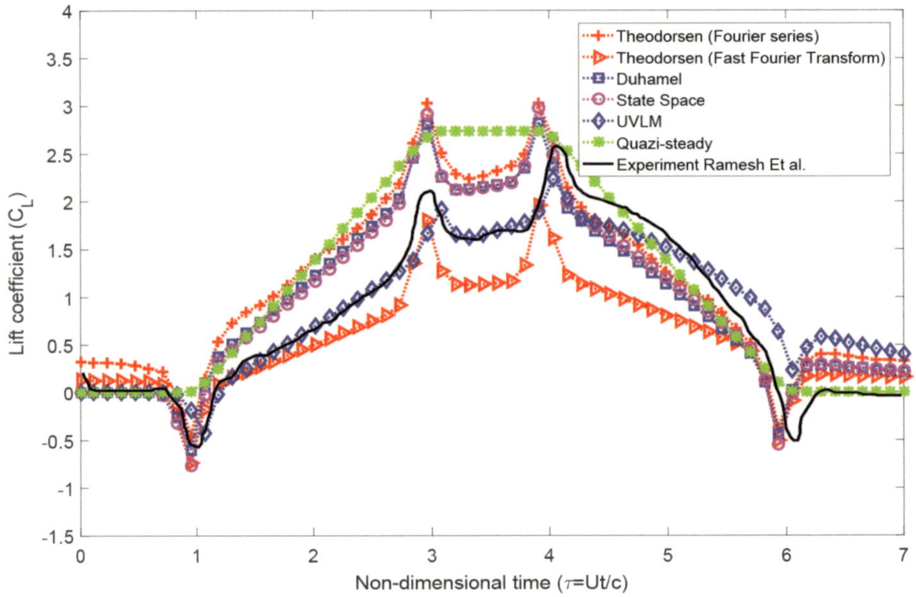

Figure 8.
Comparison for the proposed models and experimental work done by Ramesh et al. with ramp rate of 0.2 and amplitude 25° at the leading edge hinge location.

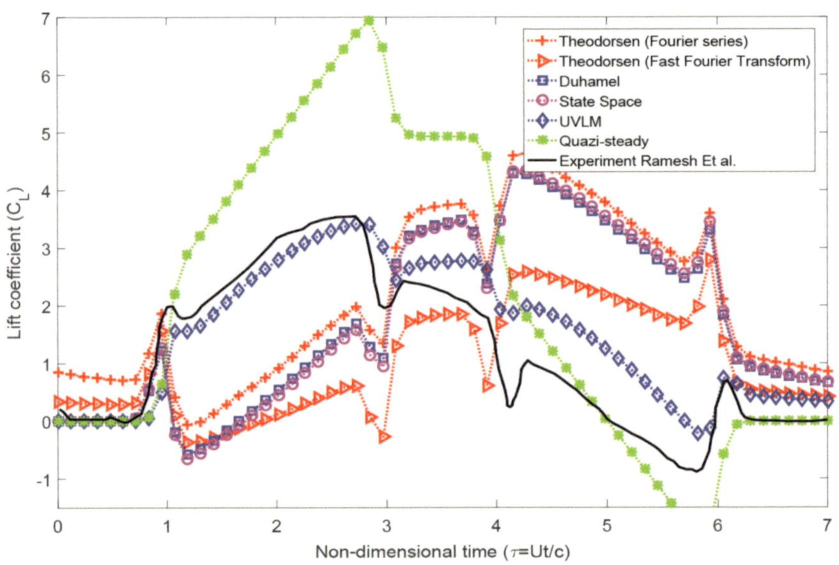

Figure 9.
Comparison for the proposed models and experimental work done by Ramesh et al. with ramp rate of 0.4 and amplitude 45° at leading pivot location.

(attenuated response) and the quasi-steady model (over predicted). Of particular interest, Duhamel and state space models coincide on top of each other having the same lift magnitude. This is excepted as the two models have the same mathematical base. At the ramp hold phase, the value of the saturated lift coefficient is approximately 2 for all models except the quasi-steady and Theodorsen based on FFT. Here, the impact of shifting the pivot location towards the trailing edge conducive in reducing the rotational effects which in turn decreases the lift coefficient by 15% at the ramp-up event compared to the quarter chord location. This

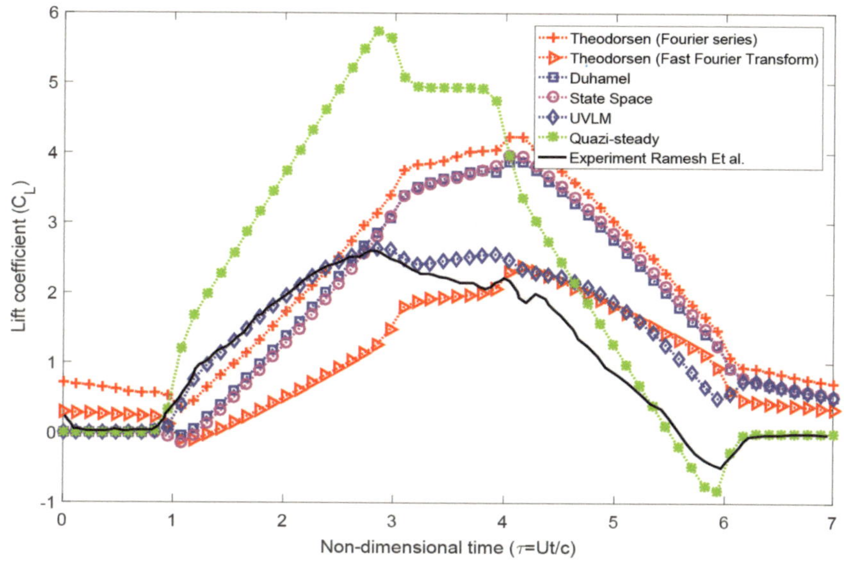

Figure 10.
Comparison for the proposed models and experimental work done by Ramesh et al. with ramp rate of 0.4 and amplitude 45° at half chord pivot location.

conclusion was reported also by Yu et al. [14] in their recent work for examining the effect of pivot locations on force and moment coefficients.

4.1.3 Trailing edge pivot

In a similar manner, **Figure 8** shows a comparison between experimental and theoretical predictions for 25° ramp case pitched about the trailing edge pivot. Lift coefficient comparison shows a qualitatively good agreement between the experiment and all the presented models (lift coefficient pattern). Of particular interest, taking the lift transition peaks during different ramp regimes. UVLM model results match very well for the ramp-up and ramp-hold regimes then decrease slightly at the ramp-down regime to give an damped lift coefficient values. Theodorsen model based on Fast Fourier Transform records an attenuated lift coefficient compared to experiments. On the other hand, all other presented models show an over predicted lift coefficient compared to the experimental results while preserving the same lift response pattern for all ramp regimes.

The common result in all pivot location cases (leading, half and trailing chord location), show that Theodorsen FFT model has a damped lift response compared to all the proposed models and experiments. This is because for a given AOA (α), one could be interested in the transient response, however, the analytical expressions cannot be obtained and the discrete Fourier Transform (FFT) has to be used instead. Discrete Fourier Transform compared to the exact Fourier Transform ignores some frequency content mathematically. In addition to the aside notion of flow dynamics, the leading edge location experiences rich dynamics when compared to the half and trailing pivot chord locations reported by Ford and Babinsky [41]. This is due to the inclusion of rotational effects which increase by increasing the spacing between the hinge point and the three-quarter chord location. It is clear that all models capture the lift peaks during transition between pitch ramps except the quasi-steady model. This lift peak has been reported by previous studies due to a delay in stall and/or a delay in LEV formation [39, 42].

4.2 Case 1: Pitch ramp $\alpha_o = 45^o$

4.2.1 Leading edge pivot

Figures 9–11 show the lift coefficient response for a ramp maneuver with an amplitude of 45° at three different pivot locations. At the leading edge pivot location; shown in **Figure 8**, at the beginning of the ramp ($\tau = 0 - 1$), Theodorsen model based on Fourier series model shows higher lift coefficient than all the other models as well as the experimental results. For the ramp upstroke, all the models showed a decrease in the lift coefficient compared to the experiment results preserving the same slope until the start of the second event then a continuous increase in lift response which appears as over predicted values compared to the experimental results presented by Ramesh et al. [9]. The UVLM model pertained the same lift pattern and all proposed models show a large discrepancy compared to experimental results. In addition, the UVLM model results show a good agreement with experiments at the ramp-up then starts to deviate with an increase in lift coefficient by 48% at hold-on and ramp down regimes. Furthermore, the quasi steady model shows a high lift coefficient at the end of the ramp-upstroke compared to the experiments, followed by a sharp decrease at the ramp-down stroke.

4.2.2 Half chord pivot

Figure 10 shows the lift coefficient for a ramp amplitude of 45° at half chord pivot location. The proposed models show a good match at the ramp-up regime then an over predicted lift coefficient occurs after the ramp-hold and ramp-down regimes compared to the experimental results except for Theodorsen FFT based model and the quasi-steady model. Again, Theodorsen FFT based model gives an attenuated response, and the quasi-steady model shows a magnified response (qualitatively similar to the results presented in **Figure 7** of a pitch ramp amplitude of 25°). During the ramp-hold phase, the UVLM model matches well with small discrepancy compared to all other models. At the final phase (ramp-down), all

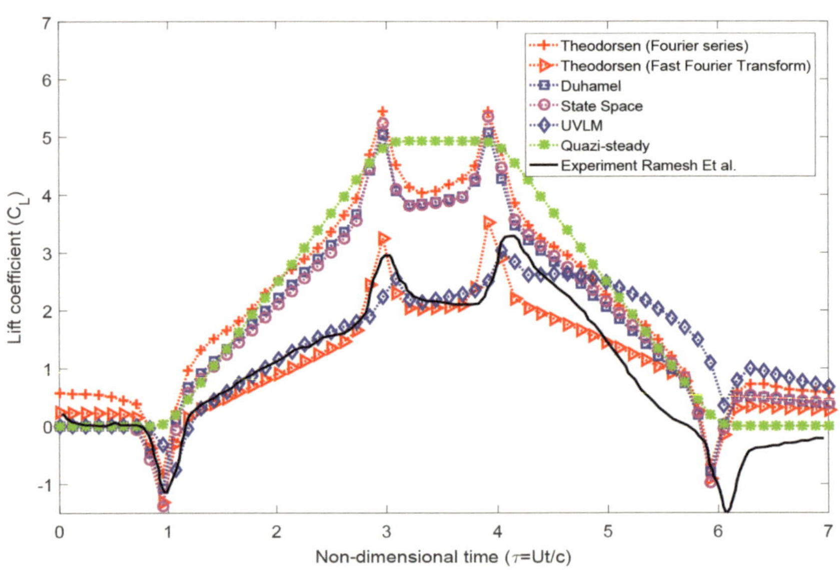

Figure 11.
Comparison for the proposed models and experimental work done by Ramesh et al. with ramp rate of 0.4 and amplitude 45° at trailing pivot location.

models show an over prediction for the lift coefficient. The results show a smooth transition without any sharp peaks in lift coefficient between different events.

4.2.3 Trailing edge pivot

Figure 11 presents the lift coefficient for a ramp amplitude of 45° at trailing edge pivot location. By comparing **Figure 11** along with **Figure 8**, the two models (Thoedorsen FFT based and UVLM) show a very good prediction with the experimental results for the two phases (ramp-up and ramp hold), then show an increase in the lift coefficient at ramp down. On the contrary, all other models record an over predicted lift coefficient compared to the experimental results for all events preserving the lift response pattern. The quasi steady models for the two ramp cases (0°-25°-0° and 0°-45°-0°) at the same pivot location (trailing edge). **Figure 8** and **Figure 11**, do not show any sharp peak for lift coefficient for the ramp transition regimes. This is expected due to the lack of inclusion of wing stall and rotational effects.

It is clear that a very good matching found between the UVLM model and experiments which can be attributed to the favor of leading edge suction inclusion as well as the nonlinear behavior ($\sin(\alpha)$) that is induced by the no-penetration boundary condition in the UVLM model. Consequently, at this range of AoA (25°) (attached flow), the dominant effect for the LES and nonlinearity associated with the ramp maneuver appears to be matched well with the results of Ramesh et al. [9]. At high angle of attack maneuver (45°), this effect no longer exists as the flow separates and became more pronounced [43]. Recall that rotational lift is proportional to the distance between the pivot and three quarter chord point (Giacomelli and Pistolesi theorem [44]), which attains and preserves its largest value for a leading edge pivot. The UVLM results match the experimental results with small

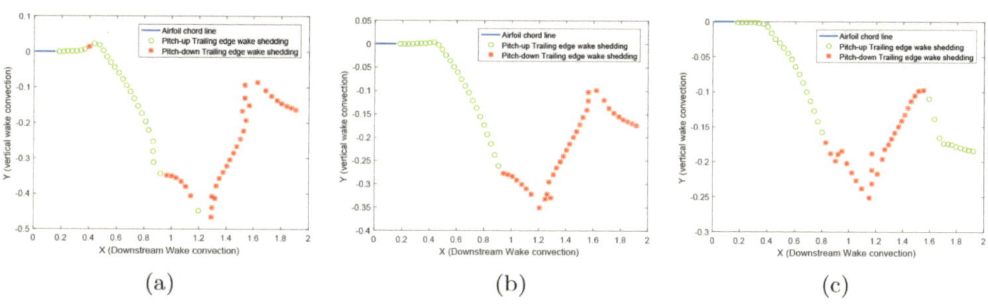

Figure 12.
Shedding of trailing vortices and wake convection downstream for 25° amplitude ramp maneuver. (a) Leading edge pivot. (b) Half chord pivot. (c) Trailing edge pivot.

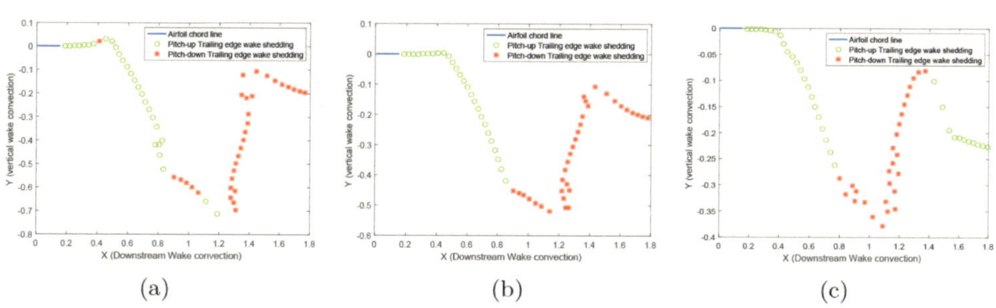

Figure 13.
Shedding of trailing vortices and wake convection downstream for 45° amplitude ramp maneuver. (a) Leading edge pivot. (b) Half chord pivot. (c) Trailing edge pivot.

nuances even for large amplitude (45°) at the ramp-up regime and partially at the ramp-hold only. At ramp-down regime, the UVLM results deviate from the experimental results and appeared to be over predicted.

Figures 12 and **13** show the Shedding of trailing vortices and wake convection shape downstream for 25° and 45° amplitudes ramp maneuvers. All the figures show the same convection pattern for the three pivot locations with an increase in the y axis vortex location with increasing the ramp amplitude.

5. Conclusion

In this chapter, different classical analytical models were presented in a simple mathematical form based on potential flow to solve unsteady problems constrained by an input motion. A canonical pitch ramp motion is chosen to present the input motion for two different ramp amplitudes (25° and 45°) and three pivot location on the airfoil chord $(c/4, c/2, 3c/4)$. The analytical results were compared to the experimental data and the comparison revealed an acceptable agreement at the pitch ramp amplitude of 25° compared to the results presented by the 45° ramp amplitude case. Thus, those models can be considered as promising aerodynamic models for predicting lift coefficient for such manoeuver at a ramp amplitude up to 25° only. Along the four analytical models, the UVLM showed very good results for the two ramp amplitude cases. It should be noted that, the UVLM captures all geometric nonlinearities, wake deformation, rolling wake, leading edge suction and post stall without the inclusion of leading edge vortex effects. Duhamel and the state space models appear to have the same behavior which asserts that the state space model shares the same physical base and obtained the same results compared to Theodorsen's model.

Table 2 discuses and concludes the output of each proposed model with the perspective of output response, pitch amplitudes, computational cost and the obtained loads.

The benefits of the UVLM compared to other methods is that is enabling aerodynamic modeling for arbitrary motion. An extension is easy to implement to include a formulation of the boundary conditions for arbitrary three-dimensional motion and control surface rotation. Furthermore the calculation of unsteady induced drag by a nonlinear extension of the force computation can be done. Furthermore the proposed UVLM method shows advantages in predicting unsteady aerodynamic forces of high frequency motion compared to other analytical models. In general, it can be said that the unsteady vortex lattice method is a powerful tool for modeling of incompressible and inviscid unsteady aerodynamics. A continuous time formulation in particular can be used to decrease the computational costs for aeroelastic simulations. The possibility of calculating unsteady loads without the need of approximations for time-domain simulation makes the method especially useful within aeroservoelastic optimization algorithms. Other models formulated in

Models	Response type	Large amplitude	Computational cost	Loads
Theodorsen	Steady state harmonic	×	×	force
Wagner	Transient	×	×	Force
State space	Full response	×	×	Force
UVLM	Full response	√	√	Pressure

Table 2.
Proposed models output parameters for solving pitching maneuvers.

time domain (for example sensor and actuator models or control laws) can be easily integrated. Furthermore, the nonlinear aerodynamic state space formulation is suitable for the integration of further nonlinear aerodynamic correction models (e.g. stall models). This provides confidence towards the development of semi-empirical models based on potential flow theories and experiments that can predict unsteady forces of ramp maneuvers.

Nomenclature

b	airfoil semi-chord $(c/2)$
c	airfoil semi-chord $(c/2)$
C_L	lift coefficient
$C(k)$	lift deficiency factor
f	frequency (Hz)
h	plunging displacement (mm)
$\dot{h}$	plunging velocity
$\ddot{h}$	plunging acceleration
k	reduced frequency $\pi fc/U_\infty$
ℓ	wing span (m)
P	non-dimensional Laplace operator
q	non-dimensional pitch rate,$\frac{\dot{\alpha}c}{V}$
Re	Reynolds number
S	distance traveled in semi-chords,$\frac{2Vt}{c}$
T	time period
U_∞	free stream velocity
U_{rel}	free stream velocity
α_o	airfoil mean angle of attack
α_{eff}	effective angle of attack
$\dot{\alpha}$	angular pitch velocity (rad/s)
$\ddot{\alpha}$	angular pitch acceleration (rad/s^2)

Greek variables

ϕ	phase angle
Γ	total flow circulation
γ_b	elementary bound flow circulation
γ_w	elementary wake flow circulation
ω	angular frequency,(rad/s)
σ	heaviside function variable
τ	Non-dimensional time
ρ	Air density

Abbreviations

AoA	angle of attack
circ	circulatory
FFT	Fast Fourier Transform
RHS	right hand side
UVLM	unsteady vortex lattice method

Author details

Mohamed Yehia Zakaria
Aircraft Mechanics Department, Military Technical College, Cairo, Egypt

*Address all correspondence to: zakaria@mtc.edu.eg

References

[1] Cory, R. and Tedrake, R. (2008). "Experiments in fixed-wing uav perching." *Proceedings of the AIAA Guidance, Navigation, and Control Conference*, AIAA Reston, VA, 1–12.

[2] Hoburg, W. and Tedrake, R. (2009). "System identification of post stall aerodynamics for uav perching." *Proceedings of the AIAA Infotech@ Aerospace Conference*, 1–9. https://en.wikipedia.org/wiki/Swallow, <https://en.wikipedia.org/wiki/Swallow>.

[3] Katz, J. and MASKEW, R. (1988). "Unsteady low-speed aerodynamic model for complete aircraft configurations." *Journal of Aircraft*, 25 (4), 302–310.

[4] Schlichting, H. T. and Truckenbrodt, E. A. (1979). *Aerodynamics of the Airplane*. McGraw-Hill Companies.

[5] Theodorsen, T. and Mutchler, W. (1935). "General theory of aerodynamic instability and the mechanism of flutter.

[6] Yan, Z., Taha, H. E., and Hajj, M. R. (2014). "Geometrically-exact unsteady model for airfoils undergoing large amplitude maneuvers." *Aerospace Science and Technology*.

[7] Theodorsen, T. (1935). "General theory of aerodynamic instability and the mechanism of flutter." *Report No. 496*, NACA.

[8] Zakaria, M. Y., dos Santos, C. R., Dayhoum, A., Marques, F. D., and Hajj, M. R. (2019). "Modeling and prediction of aerodynamic characteristics of free fall rotating wing based on experiments." *IOP Conference Series: Materials Science and Engineering*, 610, 012098.

[9] Roderick, W. R., Cutkosky, M. R., and Lentink, D. (2017). "Touchdown to take-off: at the interface of flight and surface locomotion." *Interface Focus*, 7 (1), 20160094.

[10] Hammer, P., Altman, A., and Eastep, F. (2014). "Validation of a discrete vortex method for low reynolds number unsteady flows." *AIAA journal*, 52(3), 643–649.

[11] Hammer, P., Altman, A., and Eastep, F. (2011). "A discrete vortex method application to high angle of attack maneuvers." *29th AIAA Applied Aerodynamics Conference*, 3007.

[12] Eldredge, J. D., Wang, C., and Ol, M. (2009). "A computational study of a canonical pitch-up, pitch-down wing maneuver." *AIAA paper*, 3687, 2009.

[13] Ramesh, K., Gopalarathnam, A., Edwards, J. R., Ol, M. V., and Granlund, K. (2013). "An unsteady airfoil theory applied to pitching motions validated against experiment and computation.." *Theoretical and Computational Fluid Dynamics*, 1–22.

[14] Yu, Y., Amandolese, X., Fan, C., and Liu, Y. (2018). "Experimental study and modelling of unsteady aerodynamic forces and moment on at plate in high amplitude pitch ramp motion." *Journal of Fluid Mechanics*, 846, 82120.

[15] Pitt Ford, C. and Babinsky, H. (2013). "Lift and the leading-edge vortex." *Journal of Fluid Mechanics*, 720, 280–313.

[16] Shehata, H., Zakaria, M. Y., Hajj, M. R., and Woolsey, C. A. (2019). "Aerodynamic response of a naca-0012 airfoil undergoing non-sinusoidal pitching waveforms." *AIAA Scitech 2019 Forum*, 0303.

[17] Sheng, W., Galbraith, R. M., and Coton, F. (2006). "A new stall-onset criterion for low speed dynamic-stall."

Journal of Solar Energy Engineering, 128 (4), 461–471.

[18] STRGANAC, T. and MOOK, D. (1986). “Application of the unsteady vortex-lattice method to the nonlinear two-degree-of-freedom aeroelastic equations." *27th Structures, Structural Dynamics and Materials Conference*, 867.

[19] Zakaria, M. Y., Al-Haik, M. Y., and Hajj, M. R. (2015). “Experimental analysis of energy harvesting from self-induced flutter of a composite beam." *Applied Physics Letters*, 107(2), 023901.

[20] Zakaria, M. Y. (2018). “Low to medium fidelity models for unsteady pitching maneuvers at low reynolds number." *2018 AIAA Aerospace Sciences Meeting*, 0040.

[21] Dayhoum, A., Zakaria, M. Y., and E. Abdelhamid, O. (2020). “Elastic torsion effects on helicopter rotor loading in forward flight." *AIAA Scitech 2020 Forum*, 0507.

[22] (2019). *Unsteady Aerodynamic Modeling and Prediction of Loads for Rotary Wings in Forward Flight*, Vol. Volume 8: 31st Conference on Mechanical Vibration and Noise of *International Design Engineering Technical Conferences and Computers and Information in Engineering Conference*. V008T10A047.

[23] Michael, V., Altman, A., Eldredge, J. D., Garmann, D. J., and Lian, Y. (2010). “Resume of the aiaa fdtc low reynolds number discussion groups canonical cases.

[24] dos Santos, C. R., Q. Pacheco, D. R., Taha, H. E., and Zakaria, M. Y. (2020). “Nonlinear modeling of electro-aeroelastic dynamics of composite beams with piezoelectric coupling." *Composite Structures*, 112968.

[25] Mook, D. T. and Nayfeh, A. H. (1985). “Application of the vortex-lattice method to high-angle-of-attack subsonic aerodynamics." *SAE Transactions*, 517–532.

[26] Shehata, H., Zakaria, M., Hussein, A., and Hajj, M. R. (2018). “Aerodynamic analysis of flapped airfoil at high angles of attack." *2018 AIAA Aerospace Sciences Meeting*, 0037.

[27] Van der Wall, B. and Leishman, J. G. (1994). “The influence of variable flow velocity on unsteady airfoil behavior." *J. American Helicopter Soc.*, 39(4).

[28] Tietjens, O. K. G. and Prandtl, L. (1957). *Applied hydro-and aeromechanics: based on lectures of L. Prandtl*. Courier Dover Publications.

[29] Garrick, I. E. (1938). “On some reciprocal relations in the theory of nonstationary flows.

[30] Leishman, J. G. and Beddoes, T. S. (1989). “A semi-empirical model for dynamic stall." *J. the American Helicopter Soc.*, 34(3), 3–17.

[31] Marques, F. D., Pereira, D. A., Zakaria, M. Y., and Hajj, M. R. (2017). “Power extraction from stall-induced oscillations of an airfoil." *Journal of Intelligent Material Systems and Structures*, 1045389X17739161.

[32] Wagner, H. (1925). “Uber die entstehung des dynamischen auftriebes von tragugeln." *Zeitschrift fur Angewandte Mathematic und Mechanik*, 35, 17.

[33] M Y Zakaria, D. A. Periera, S. R. M. h. M. D. (2017). “An experimental assessment of unsteady forces on a plunging airfoil oscillating in stationary fluid at high frequencies." *Engineering of Science and Military Technologies Journal*, 1, 13.

[34] M.Y. Zakaria, H.E. Taha, M. H. (2017). “Measurement and modeling of lift enhancement on plunging airfoils: A

frequency response approach." *Journal of Fluids and Structures*, 69(February 2017), Pages–187.

[35] Taha, H. E., Hajj, M. R., and Beran, P. S. (2014). "State-space representation of the unsteady aerodynamics of flapping flight." *Aerospace Science and Technology*, 34, 1–11.

[36] Kussner, H. and Schwartz, I. (1941). "The oscillating wing with aerodynamically balanced elevator.

[37] Katz, J. and Plotkin, A. (2001). *Low Speed Aerodynamics*. Cambridge University Press, 2 edition.

[38] Binder, S., Wildschek, A., and De Breuker, R. "Extension of the continuous time unsteady vortex lattice method for arbitrary motion, control surface deection and induced drag calculation.

[39] Sodja, J., Drazumeric, R., Kosel, T., and Marzocca, P. (2014). "Design of flexible propellers with optimized load-distribution characteristics." *Journal of Aircraft*, 51(1), 117–128.

[40] Konstadinopoulos, P., Thrasher, D., Mook, D., Nayfeh, A., and Watson, L. (1985). "A vortex-lattice method for general, unsteady aerodynamics." *Journal of aircraft*, 22(1), 43–49.

[41] Ramesh, K., Gopalarathnam, A., Edwards, J. R., Ol, M. V., and Granlund, K. (2011). "Theoretical, computational and experimental studies of a at plate undergoing high-amplitude pitching motion." *AIAA Paper*, 217.

[42] Granlund, K. O., Ol, M. V., and Bernal, L. P. (2013). "Unsteady pitching at plates." *Journal of Fluid Mechanics*, 733.

[43] Yu, H.-T., Bernal, L. P., and Morrison, C. (2012). "Experimental investigation of pitch ramp-hold-return motion of at plates at low reynolds number." *AIAA Paper*, 51.

[44] Giacomelli, R. and Pistolesi, E. (1934). "Aerodynamic theory." *Div. D. (ed. Wm. F. Durand). Berlin: Springer*.

Chapter 7

An Introduction of Biomimetic System and Heat Pump Technology in Food Drying Industry

Khurram Yousaf, Kunjie Chen and Muhammad Azam Khan

Abstract

Drying of food products is a relatively complex, nonlinear, and dynamic process due to simultaneous heat and mass transfer, rapid moisture evaporation, and biological and chemical reactions. Therefore, the monitoring of food quality during the drying process using bio-inspired technologies can play a vital role. The demand for high-quality dried food products and the rapid growth of energy in food processing are attracting new and renewable sources of energy. Energy efficiency, improved food product quality, and less environmental impact are always the main priorities of any drying system development. In-depth knowledge of biomimetic systems and drying kinetics would be helpful to design new dryers and technologies. Due to the excellent features (controllable drying temperature, drying time, drying air velocity, and relative humidity), heat pump drying systems have been used widely to ensure food and agricultural product quality. This chapter helps to understand the relationship between bio-inspired technologies and the role of heat pump technology in the food drying industry in terms of cost-effectiveness, energy saving, and better food product quality.

Keywords: heat pump, drying, biomimetic, food quality, energy efficiency

1. Introduction

Drying remains an essential part of food processing to enhance the shelf life of agricultural produce. To facilitate storage, preserve nutritional value, and add value in farm products, drying has been an essential solution [1]. Vegetables and fruits are one of the primary sources of vitamins and minerals. They have a high amount of nutritional components due to which they are considered a basic need for a balanced diet. Food products contained a high amount of moisture, such as vegetables and fruits, which have more than 80% and cereals more and less 30% [2]. Therefore, vast amounts of foodstuff deteriorate annually because of the inappropriate handling (pre- and postharvesting) and lack of processing facilities. Approximately 30–50% of losses occur in vegetables and fruits from field to consumer. The easy availability of fresh food to consumers for an extended period, process, and storage facilities are compulsory [3]. However, the increasing infrastructure and processing cost of post-storage techniques are essential limitations. Due to increasing demand of high-quality food products, various conventional and mechanical food drying methods have been employed to process and protect from deterioration such as solar drying, oven drying, hot air drying, vacuum drying, freeze drying, microwave drying, infrared drying, and air source heat pump (ASHP) drying [4].

The ASHP drying system has two main components: an HP and a drying assembly—a functional HP system is composed of a compressor, condenser, evaporator, refrigerant, and throttle valve [5, 6]. Similarly, the drying system consists of the dryer and other necessary accessories. In the drying cycle, the materials to be dried are fed into the drying chamber and dehydrated until attained desired moisture content. In the HP cycle, the condenser provides the heat for the drying process by condensation of the refrigerant; similarly, the refrigerant absorbs heat and evaporates in the evaporator. The compressor of the HP compressed it again and delivered to the condenser to complete the cycle [7]. The hot heat generated by the condenser and cold heat by the evaporator can be used at the same time during the HP operation. The hot and cold heat from condenser and evaporator will be used to heat the product and dehumanization, respectively. Due to the excellent capability of the HP to convert the latent heat of vapor condensation into sensible heat of air makes it attractive in the drying industry. Many investigators recognized the utilization of HP in drying, which has attracted applied and theoretical research on drying using the HP [8–10].

The energy-saving capability and the ability to control air humidity and temperature are the main advantages of HP technology. These factors create options for variable drying conditions. Drying is one of the most energy-intensive practices, as it consumes up to 15% of industrial energy and 9–25% of national energy in developed countries [11]. About 40% of energy could be saved by using HP dryers as compared to electric resistance dryers. The drying process utilizes up to 70, 50, and 60% of total energy in the manufacturing of wood products, textile fabrics, and farm corn production, respectively [12]. Therefore, it is necessary to find out appropriate HP technology to reduce energy consumption. It has been reported that HP drying helps to improve product quality.

The performance of the HP system can be improved by the implementation of a sophisticated control strategy [13]. The three-layer ANN model has been used in the prediction of HP system performance. The study depicted that the application of ANN in the prediction of the HP system was consistent and robust [14]. Yang et al. [15] have used a simultaneous control strategy to improve the precision of drying temperature and superheat. A stable drying temperature was attained using two fuzzy controllers with different membership functions and control rules. The study revealed that the newly developed controller is reliable to reduce the nonlinearities of drying temperature and superheat. The yum slices of different thicknesses and longan were dried using the step-down and continues dehumidification strategies of controlling relative humidity [16]. It concluded that a step-down approach was found more energy efficient for 12 or 18 mm thickness, whereas a continuous decrease of relative humidity strategy was found energy efficient for longan, respectively.

Besides these drying systems, smart drying mechanisms (biomimetic systems) can be cost-effective in controlling various foodstuffs quality parameters with variable drying process time. The quality of food products can be assessed through many senses such as appearance, smell, and taste. The human sense inspired systems including electronic noses (odor sensing technology) and electronic tongues (taste sensing technology) are extensively under considerations by researchers and scientists for different food processing steps, e.g., drying, packaging, and storage. These biomimetic systems provide reasonable information on odor and taste to analyze the quality of food products. The combination of these biomimetic systems, computer software, and data analysis methods exhibits environment-friendly, cost-effective, and multifunctional results for food quality control. Many studies highlighted the significance of HP technology in terms of energy efficiency, cost-effectiveness, and product quality. However, there is a need to summarize the technological advancement of the HP drying system in the food industry. The HP drying system can be more sophisticated and attractive if coupled with biomimetic odor and taste sensing systems for food safety and quality. This chapter presents the recent developments in HP drying technologies of food products, studies these advancements to maintain the food quality, and further discusses the research potential in this field.

2. HP drying and its application in the food industry

2.1 HP drying

Heat pumps raise the energy gained from the low-temperature level to high-temperature level and transfer it to the energy carrier medium. In an HP, both the heating and cooling processes of the refrigerator are utilized. Nowadays, HP technology is used in the domestic and commercial sectors for space heating and cooling such as in offices, water heating, swimming pools, commercial drying in agriculture, and the wood industry as well as the cotton industry. The main objective of the utilization of HP technology in the food drying sector is to dehydrate the food products for quality enhancement at minimum cost. HP drying technology is one of the most efficient and controllable methods of water extraction from the material. The research and development in HP technology have improved its performance. HP drying technology has been modified to air source HP drying, ground source heat pump drying (GSHPD), chemical source heat pump drying (CSHPD), and hybrid heat pump drying (HHPD), as shown in **Figure 1**. The following chapter reflects the contribution of significant HP technologies in the food drying sector.

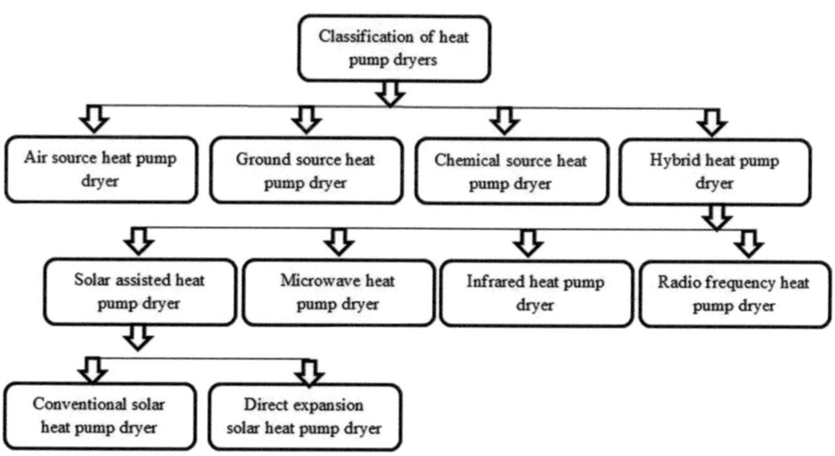

Figure 1.
The primary classification of heat pump dryers [17].

2.2 Performance indicators of HP drying system

2.2.1 The coefficient of performance

The coefficient of performance (COP) of the HP system refers to estimate the performance during heating conditions. In an ideal model, the heating of hot air in the drying chamber is a hot air adiabatic cooling process as follows:

$$c_p\left(T_{dr,a,in}-T_{dr,a,out}\right)=\left(d_{dr,a,out}-d_{dr,a,in}\right)r_s \tag{1}$$

In actual condition, enthalpy change occurs during hot air passed through the drying chamber as follows:

$$\Delta h_{dr,\,a}=c_p\left(T_{dr,\,a,\,in}-T_{dr,a,\,out}\right)-r_s\left(d_{dr,\,a,\,out}-d_{dr,a,\,in}\right) \tag{2}$$

For materials,

$$DR=\frac{M_t-M_{t-\Delta t}}{\Delta t} \tag{3}$$

During the hot air cycle, two heat exchange processes occur in the HP drying system, as follows:

$$Q_{evap}=m_{evap,a}C_{p,a}\left(T_{evap,a,in}-T_{evap,a,out}\right) \tag{4}$$

$$Q_{cond}=m_{cond,a}C_{p,a}\left(T_{cond,a,out}-T_{cond,a,in}\right) \tag{5}$$

where

$$C_{p,a}=1.01+1.84d \tag{6}$$

$$W_h=m_a\left(h_{dr,a,in}-h_{HP,a,in}\right) \tag{7}$$

$$W_{comp}+W_h=m_a\Delta h_{dr,a}+\Delta Q_{HP} \tag{8}$$

$$W_{comp}+W_h=m_a\left(\Delta h_{dr,\,a}+h_{HP,a,out}-h_{HP,a,in}\right) \tag{9}$$

The compressor power and speed can be controlled by motor frequency. The COP of the HP and the whole system can be measured using the following relationships:

$$COP_{hp}=\frac{Q_{cond}}{W_{copm}} \tag{10}$$

$$COP_S = \frac{Q_{cond}}{\sum W} \tag{11}$$

2.2.2 Specific moisture evaporation rate

The ratio of water removed from material to total energy consumption is termed a specific moisture evaporation rate (SMER). The HP drying system must apply energy and dehumidified material. It must be measured to evaluate the performance of the heat pump drying system. The SMER (kg/kWh) is given as follows:

$$SMER = \frac{M_{evp}}{W_h + W_f + W_{comp}} \tag{12}$$

2.2.3 Moisture extraction rate

It is defined as water extracted from material to the total time of material drying in the chamber and expressed in kg/h. Moisture extraction rate (MER) is one of the important factors to evaluate the HP drying system performance. MER can be calculated as follows:

$$MER = \frac{M_{evp}}{t_{dr}} \tag{13}$$

2.3 Air source heat pump drying of food

An air source HP dryer works on a simple principle of HP technology with the condenser as a heat source and evaporator as a dehumidifier. Liu et al. [6] designed a multifunctional air source HP dryer and investigated the drying performance of 3-mm thickness garlic slices in three different ambient temperature and humidity conditions. The drying rate and energy consumption of the HP drying system were investigated by utilizing different hot air circulation modes using an enthalpy-humidity diagram.

An air source HP drying system not just depends on the HP but also the circulation style of the drying medium. According to the circulation style and degree of ventilation, ASHP dryer can be divided into open, semi-open, and closed types. The exhaust gas flowing out of the drying chamber contains a high amount of heat; the temperature of exiting gas is usually higher as compared to the ambient temperature. As a result, it is essential to pass the exiting gas through the evaporator of the HP to increase the thermal efficiency of the HP drying system. Sometimes, it restricts the heat exchange due to dust particles adhered to the surface of the evaporator. BAR can be adjusted by changing the degree of opening and closing of duct valves. The BAR can be expressed as follows:

$$BAR = \frac{a_{bypass}}{a_{through\ evaporator}} \tag{14}$$

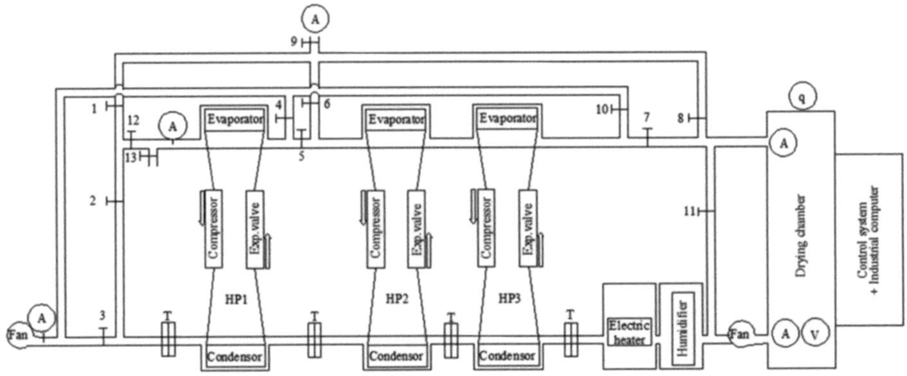

Figure 2.
Air source heat pump dryer test bench with a series of heat pump units [5]. (A = temperature and humidity sensors, q = quality sensor, V = air flow sensor, HP1, 2, 3 = heat pump units, T = temperature sensors, and 1–13 = air duct valves).

Yousaf et al. [5] used the HP dryer test bench with a series of heat pumps for parboiled rice drying of different varieties and sample loads, as shown in **Figure 2**. SMER, MER, and COP of HP dryer were calculated and analyzed in open and closed cycles. The COP of the closed-loop cycle (2.211) drying was higher than open-loop cycle (1.409) because fresh air entered into the system during the open-loop cycle. Saensabai and Prasertsan [18] conducted a simulation study of five different configurations of the air source HP. Soponronnarit et al. [19] reported that COP_{hp} and SMER enhanced with 0% BAR and reduced with increasing of BAR during paddy seed drying. Chapchaimoh et al. [20] utilized the closed system HP dryer for ginger drying, where air and nitrogen were used as a drying medium. The study concluded that SMER in air and nitrogen was 0.06 and 0.07 kg H_2O/MJ, respectively. Similarly, energy consumption for air and nitrogen drying was 16.67 and 14.29 MJ/kg, respectively. Many studies have been done on drying of food products using different HP technology systems, as shown in **Table 1**.

2.4 Ground source or geothermal heat pump drying of food

Recent developments in HP technology made it attractive for researchers to conserve energy and processing time. The geothermal HP system can save 44 and 70% energy as compared to air source heat pumps and air conditioning and conventional electrical heating [54]. The ground source heat pump (GSHP) drying system is relatively similar to the ASHP drying system with the addition of a ground source heat exchanger, as shown in **Figure 3**. The following system extracts heat from the ground by a ground source heat exchanger with the help of a circulated antifreeze water solution. The extracted heat is transferred to refrigerant in the evaporator and delivered to the food drying chamber. However, few studies on GSHP drying of food have been reported.

Colak et al. [29] designed a GSHP dryer unit for a single layer of mint leaves drying. The drying system was designed and developed in Solar Energy Institute, Ege University, Izmir, Turkey. The mint leaves were dried at three different drying temperatures 40, 45, and 50°C, constant relative humidity of 16%, and mass flow rate range of 0.01–0.05 kg/s. It revealed that the highest exergy efficiency (97.24%) was achieved at maximum temperature (50°C) and a mass flow rate of 0.05 kg/s. Another researcher studied the energy and exergy analysis of the GSHP drying

Sr. no.	Researcher	Food product	Heat pump dryer type	Reference
1	Achariyaviriya et al.	Papaya glace	Air source heat pump	[21]
2	Aktas et al.	Apple	Vapor compression cycle	[22]
3	Artnaseaw et al.	Chili	Vacuum heat pump	[23]
4	Artnaseaw et al.	Shiitake mushroom	Vacuum heat pump	[24]
5	Best et al.	Rice	Solar-assisted vapor compression cycle	[25]
6	Chen et al.	Lemon	Vapor compression cycle	[26]
7	Chua et al.	Guava and banana	Two-stage heat pump	[27]
8	Colak and Hepbasli	Apple	Ground source heat pump	[28]
9	Colak et al.	Mint leaves	Ground source heat pump	[29]
10	Cunney and Williams	Grain	Air source heat pump	[30]
11	Fatouh et al.	Jew's mallow, spearmint, and parsley	Air source heat pump	[31]
12	Hancioglu and Hepbasli	Laurel leaves	Ground source heat pump	[32]
13	Hawlader et al.	Yam	Air source heat pump	[33]
14	Hawlader et al.	Food grains	Air source heat pump	[34]
15	Hawlader and Jahangeer	Green beans	Air source heat pump	[35]
16	Hawlader et al.	Apple and guava	Dual condenser vapor compression cycle	[36]
17	Ho et al.	Potatoes	Air source heat pump	[37]
18	Icier et al.	Agriculture broccoli floret	Dual condenser vapor compression cycle	[38]
19	Icier and Erbay	Olive leaf	Dual condenser vapor compression cycle	[39]
20	Jia et al.	Foam rubber, sliced carrots, and root	Air source heat pump	[40]
21	Pal et al.	Green sweet pepper	Dual condenser vapor cycle	[41]
22	Prasertsan and Saen-saby	Red pepper	Atmospheric freezer heat pump	[42]
23	Queiroz et al.	Tomatoes	Air source heat pump	[43]
24	Rahman et al.	Peas	Dual condenser vapor compression cycle	[44]
25	Rossi et al.	Vegetables	Air source heat pump	[45]
26	Shi et al.	Horse mackerel	Air source heat pump	[46]
27	Soponronnarit et al.	Papaya glace	Air source heat pump	[47]
28	Teeboonma et al.	Papaya and mango	Dual condenser vapor compression cycle	[48]
29	Vazquez et al.	Grapes	Vapor compression cycle	[49]
30	Zhang et al.	Carrot cubes	Air source heat pump	[50]

Sr. no.	Researcher	Food product	Heat pump dryer type	Reference
31	Aktas et al.	Mint leaves	Air source heat pump	[51]
32	Erbay and Hepbasli	Fresh laurel leaves	Ground source heat pump	[52]
33	Taşeri et al.	Grapes	Air source heat pump	[53]
34	Yousaf et al.	Paddy	Air source heat pump	[5]
35	Liu et al.	Garlic slices	Air source heat pump	[6]

Table 1.
Heat pump drying application in different food product drying.

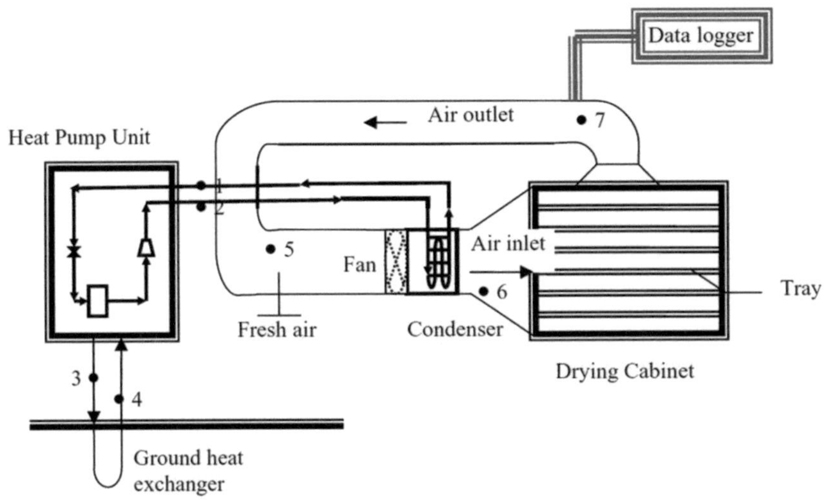

Figure 3.
Schematic diagram of ground source heat pump food drying system [29]. (1) Evaporator temperature and pressure; (2) condenser temperature and pressure; (3) fluid temperature at the ground heat exchanger inlet; (4) fluid temperature at the ground heat exchanger outlet; (5) fresh air temperature and humidity; (6) air inlet temperature and humidity; and (7) air outlet temperature and humidity.

system during laurel leaves drying. The COP of the GSHP and the whole system is in the range of 1.63–2.88 and 1.45–2.65, and SMER and SMExR of the system are in the range of 0.122 and 5: 11 kg/kWh respectively [55].

2.5 Hybrid solar-assisted heat pump drying of food

Sun drying has been used for food drying purposes for decades in tropical and subtropical countries, where solar energy is used as a primary source of energy to preserve fruits, vegetables, and other agricultural products. The traditional drying of food products by direct exposure has limitations, such as spoilage due to inappropriate climate conditions, waste of food due to the animal, birds, insects, and, most importantly, uncontrollable temperature and humidity conditions. The advancement of research and development in science and technology has changed the world; researchers and scientists are trying to find out the ways to reduce energy consumption in food processing and conservation. These globally renowned challenges in the food industry help researchers to design a hybrid drying system. Solar-assisted heat pump (SAHP) dryers are the best alternative to overcome energy crises challenges in the food industry [56]. According to literature, many studies have been done on the hybrid solar-assisted drying system, such as forced convection [57–61], solar-assisted auxiliary air heater dryer [62–64], and chemical heat pumps (CHP).

A SAHP dryer consists of a vapor compression cycle unit combined with a solar collector and drying chamber having a high coefficient of performance, as shown in **Figure 4** [65–67]. There are different kinds of SAHP dryers, depending on design, use, and material to be dried. The combination of the HP dryer and solar collector with the availability of solar energy can increase the energy efficiency and temperature of the drying system. Troger and Butler [68] studied the solar collector cum rock bed storage system for peanut drying. Gan et al. [69] studied the drying kinetics and quality of Misai Kucing leaves, stem, flowers, and mix components and compared the SAHP drying with conventional solar drying. The results were evidence of the change in color and loss of two bioactive components by solar drying. However, these changes were not seen in SAHP drying. The reason could be the higher drying temperature, long processing time, and degradation of chlorophyll.

Best et al. [70] developed a modified 7-kW air conditioning unit combined with a solar collector for rice drying. Another study of mushroom drying using a SAHP system with a flat plate solar collector and a GSHP system has been reported in the literature. Heat pump system, solar system, and SAHP system were employed for mushroom drying at 250–220, 270–165, and at 230–190 min, respectively. A computer program and PLC were used to control and monitor the relative humidity, drying air temperature, and weight of material to be dried. The mushrooms were dried from initial moisture 13.24 to 0.07 g (dry basis) using two different drying air temperatures of 45 and 55°C and 310 kg/h mass flow rate. SMER, energy utilization ratio, and COP were found to be in the range of 0.26–0.92 kg/kW h, 0.42–0.66, and 2.1–3.1, respectively [71]. Saffron drying has been studied in Iran using a solar-assisted hybrid photovoltaic thermal HP dryer. The system performance was investigated using variable drying air temperature (40, 50, and 60°C), mass flow rate (0.008, 0.012, and 0.016 kg/s), and drying modes with or without an HP. The study reported that using an HP, maximum dryer efficiency (72%) and SMER (1.16) were observed at drying air temperature of 60% and mass flow rate of 0.016 kg/s. Moreover, with the increase of drying air temperature and mass flow rate, energy consumption and drying time reduced. It also revealed, when dryer equipped with an HP, the average energy consumption decreased by up to 33% [72].

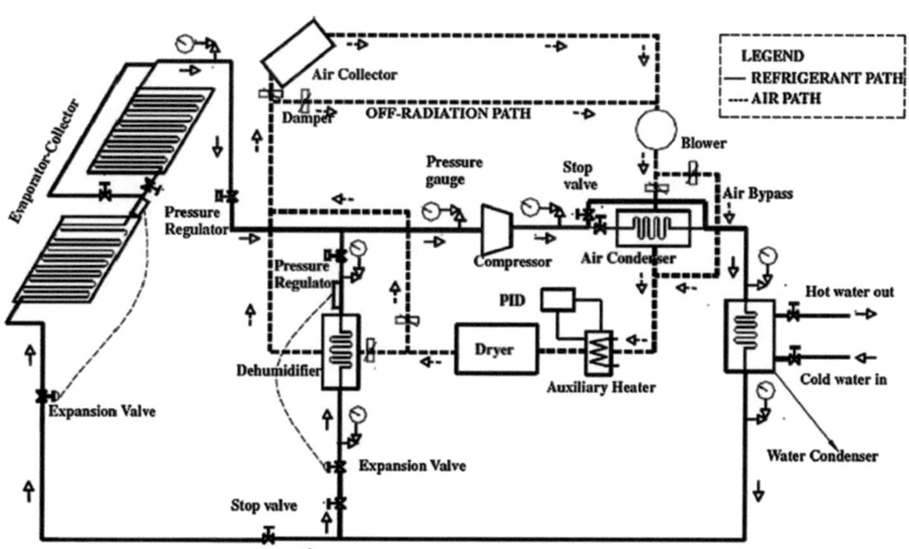

Figure 4.
Solar-assisted heat pump drying system [35].

3. Conclusion

Considering the drying kinetics of food products, HP technology is the best option for new demands of energy efficiency, product quality at a lower cost, and lower environmental impact. Many studies have discussed the HP system for food drying and concluded that HP drying systems are suitable for heat-sensitive products. The biomimetic systems are contributing as a key element in smart drying processes and becoming a very useful factor in the food drying industry. However, it has some limitations in its widespread applications in food drying industry. In the case of a bio-inspired electric nose, its sensing ability is greatly affected by humidity, background noise, temperature, sensor poisoning, and sensor variations. This chapter reported the contribution of HP technology in food drying. The demands of high-quality food products and energy-efficient technologies for food processing are increasing day by day. The future work on biomimetic systems and HP technology would allow researchers to produce more automated food drying plants involving innovative intelligent approaches. It can be seen in this book chapter, research and development in HP drying have improved to hybrid type HP dryers. No doubt, the performance of these hybrid systems has increased but would increase installation and processing costs.

Acknowledgements

The financial support of the College of Engineering, Nanjing Agricultural University, under the Jiangsu Agricultural Science and Technology Innovation Fund CX-(17)1003 is thankfully acknowledged.

Conflict of interest

The authors declare no conflict of interest.

Nomenclature

HP	heat pump
ASHP	air source heat pump
GSHP	ground source heat pump
SAHP	solar-assisted heat pump
CSHP	chemical source heat pump
BAR	bypass air ratio
SMER	specific moisture evaporation rate
MER	moisture extraction rate
ANN	artificial neural network
COP	coefficient of performance
Q	heating capacity (kW)
d	absolute humidity (kg water vapor/kg dry air)
rs	evaporative latent heat of water (kJ/kg)
m	mass flow rate (kg/s)
C_p	specific heat (kJ/kg K)
W	power (kW)
T	temperature (°C)
h	enthalpy (kJ/kg)
t, Δt	time (h)

Subscripts

a	air
evap	evaporator
f	fan
comp	compressor
s	whole system
dr	dryer, drying
cond	condenser
out	outside
in	inside

Author details

Khurram Yousaf[1,2]*, Kunjie Chen[1]* and Muhammad Azam Khan[3]

1 College of Engineering, Nanjing Agricultural University, Jiangsu, China

2 Atta-ur-Rahman School of Applied Biosciences, National University of Sciences and Technology (NUST), Islamabad, Pakistan

3 Department of Farm Machinery and Power, Faculty of Agricultural Engineering and Technology, University of Agriculture Faisalabad, Faisalabad, Pakistan

*Address all correspondence to: khurramyousaf@njau.edu.cn and kunjiechen@njau.edu.cn

References

[1] Yousaf K, Kunjie C, Cairong C, Abbas A, Huang Y, Arslan C, et al. The optimization and mathematical modeling of quality attributes of parboiled rice using a response surface method. Journal of Food Quality. 2017;**2017**(5960743):1-13

[2] Hasan MU, Malik AU, Ali S, Imtiaz A, Munir A, Amjad W, et al. Modern drying techniques in fruits and vegetables to overcome postharvest losses: A review. Journal of Food Processing & Preservation. 2019;**43**(12):e14280

[3] FAO. The State of Food and Agriculture. Rome, Italy: Food and Agriculture Organization of the United Nations; 2012

[4] Goh LJ, Othman MY, Ruslan SMH, Sopian K. Review of heat pump systems for drying application. Renewable and Sustainable Energy Reviews. 2011;**15**:4788-4796

[5] Yousaf K, Liu H, Gao X, Liu C, Abbas A, Nyalala I, et al. Influence of environmental conditions on drying efficiency and heat pump performance in closed and open loop drying of paddy. Drying Technology. 2019;**2019**(1691011):1-14

[6] Liu H, Yousaf K, Chen K, Fan R, Liu J, Soomro SA. Design and thermal analysis of an air source heat pump dryer for food drying. Sustainability. 2018;**2018**(10):1-17

[7] Liu S, Li X, Song M, Li H, Sun Z. Experimental investigation on drying performance of an existed enclosed fixed frequency air source heat pump drying system. Applied Thermal Engineering. 2018;**130**(2018):735-744

[8] Mohammadi I, Tabatabaekoloor R, Motevali A. Effect of air recirculation and heat pump on mass transfer and energy parameters in drying of kiwifruit slices. Energy. 2019;**170**:149-158

[9] Li H, Xie L, Ma Y, Zhang M, Zhao Y, Zhao X. Effects of drying methods on drying characteristics, physicochemical properties and antioxidant capacity of okra. LWT - Food Science and Technology. 2019;**101**:630-638

[10] Feng Y, Zhou C, ElGasim A, Yagoub A, Sun Y, Owusu-Ansah P, et al. Improvement of the catalytic infrared drying process and quality characteristics of the dried garlic slices by ultrasound-assisted alcohol pretreatment. LWT-Food Science and Technology. 2019;**116**:108577

[11] Chua KJ, Mujumdar AS, Hawlader MNA, Chou SK, Ho JC. Batch drying of banana pieces – Effect of stepwise change in drying air temperature on drying kinetics and product color. Foodservice Research International. 2001;**34**:721-731

[12] Mujumdar AS. Handbook of Industrial Drying. 2nd ed. Marcel Dekker Inc: New York, USA; 1987

[13] Lutfy OF, Selamat H, Noor SBM. Intelligent modeling and control of a conveyor-belt grain dryer using a simplified type-2 neuro-fuzzy controller. Drying Technology. 2015;**33**:1210-1222

[14] Wei XQ, Li NP, Peng JQ, Cheng JL, Su L, Hu JH. Analysis of the effect of the $CaCl_2$ mass fraction on the efficiency of a heat pump integrated heat-source tower using an artificial neural network model. Sustainability. 2016;**2016**(8):410

[15] Yang Z, Zhu Z, Zhao F. Simultaneous control of drying temperature and superheat for a

closed-loop heat pump dryer. Applied Thermal Engineering. 2016;**93**:571-579

[16] Ju HY, Zhao SH, Mujumdar AS, Fang XF, Gao ZJ, Zheng ZA, et al. Energy efficient improvements in hot air drying by controlling relative humidity based on Weibull and Bi-Di models. Food and Bioproducts Processing. 2018;**111**:20-29

[17] Daghigh R, Ruslan MH, Sulaiman MY, Sopian K. Review of solar assisted heat pump drying systems for agricultural and marine products. Renewable and Sustainable Energy Reviews. 2010;**14**(9):2564-2579

[18] Saensabai P, Prasertsan S. Effects of component arrangement and ambient and drying conditions on the performance of heat pump dryers. Drying Technology. 2003;**21**(1):103-127

[19] Soponronnarit S, Wetchacama S, Kanphukdee T. Seed drying using a heat pump. International Energy Journal. 2000;**1**(2):97-102

[20] Chapchaimoh K, Poomsa-ad N, Wiset L, Morris J. Thermal characteristics of heat pump dryer for ginger drying. Applied Thermal Engineering. 2016;**95**:491-498

[21] Achariyaviriya S, Soponronnarit S, Terdyothin A. Mathematical model development and simulation of heat pump fruit dryer. Drying Technology. 2000;**18**(1-2):479-491

[22] Aktas M, Ceylan I, Yilmaz S. Determination of drying characteristics of apples in a heat pump and solar dryer. Desalination. 2009;**239**:266-275

[23] Artnaseaw A, Theerakulpisut S, Benjapiyaporn C. Development of a vacuum heat pump dryer for drying chilli. Biosystems Engineering. 2009;**105**:130-138

[24] Artnaseaw A, Theerakulpisut S, Benjapiyaporn C. Drying characteristics of shiitake mushroom and Jinda chili during vacuum heat pump drying. Food and Bioproducts Processing. 2009;**88**:104-114

[25] Best R, Cruz JM, Gutierrez J, Soto W. Experimental results of a solar assisted heat pump rice drying system. Renewable Energy. 1996;**9**(1-4):690-694

[26] Chen HH, Hernandez CE, Huang TC. A study of the drying effect on lemon slices using a closed-type solar dryer. Solar Energy. 2005;**78**:97-103

[27] Chua KJ, Mujumdar AS, Chou SK, Hawlader MNA, Ho JC. Convective drying of bananas, guava, and potato pieces: Effect of cyclical variations of air temperature on drying kinetics and color change. Drying Technology. 2000;**18**(4-5):907-936

[28] Colak N, Hepbasli A. Exergy analysis of drying of apple in a heat pump dryer. In: Second International Conference of the Food Industries & Nutrition Division on Future Trends in Food Science and Nutrition. 2005. pp. 145-158

[29] Colak N, Kuzgunkaya E, Hepbasli A. Exergetic assessment of drying of mint leaves in a heat pump dryer. Journal of Food Process Engineering. 2008;**31**: 281-298

[30] Cunney MB, Williams P, editors. An engine-driven heat pump applied to grain drying and chilling. In: Proceedings Second International Symposium on the Large Scale Applications of Heat Pumps. 1984

[31] Fatouh M, Metwally MN, Helali AB, Shedid MH. Herbs drying using a heat pump dryer. Energy Conservation and Management. 2006;**47**:2629-2643

[32] Hancioglu KE, Hepbasli A. Exergetic performance assessment of a ground source heat pump drying system. International Journal of Energy Research. 2007;**31**:760-777

[33] Hawlader MNA, Bong TY, Yang Y, editors. A simulation and performance analysis of a heat pump batch dryer. In: Proceedings of the Eleventh International Drying Symposium. 1998

[34] Hawlader MNA, Chou SK, Jahangeer KA, Rahman SMA, Eugene LKW. Solar-assisted heat pump dryer and water heater. Applied Energy. 2003;**74**:185-193

[35] Hawlader MNA, Jahangeer KA. Solar heat pump drying and water heating in the tropics. Solar Energy. 2006;**80**(5):492-499

[36] Hawlader MNA, Perera CO, Tian M. Properties of modified atmosphere heat pump dried foods. Journal of Food Engineering. 2006;**74**(3):392-401

[37] Ho JC, Chou SK, Mujumdar AS, Hawlader MNA, Chua KJ. An optimization framework for drying of heat sensitive products. Applied Thermal Engineering. 2001;**21**: 1779-1798

[38] Icier F, Colak N, Erbay Z, Kuzgunkaya EH, Hepbasli A. A comparative study on exergetic performance assessment for drying of broccoli florets in three different drying systems. Drying Technology. 2010;**28**:193-204

[39] Icier F, Erbay Z. Optimization of drying of olive leaves in a pilot-scale heat pump dryer. Drying Technology. 2009;**27**:416-427

[40] Jia X, Clements S, Jolly P. Study of heat pump assisted microwave drying. Drying Technology. 1993;**11**(7): 1583-1616

[41] Pal US, Khan MK, Mohanty SN. Heat pump drying of green sweet pepper. Drying Technology. 2008;**26**: 1584-1590

[42] Prasertsan S, Saen-saby P. Heat pump drying of agricultural materials. Drying Technology. 1998;**16**(1-2):235-250

[43] Queiroz R, Gabas AL, Telis VRN. Drying kinetics of tomato by using electric resistance and heat pump dryers. Drying Technology. 2004;**22**(7): 1603-1620

[44] Rahman MS, Perera CO, Thebaud C. Desorption isotherm and heat pump drying kinetics of peas. Food Research International. 1998;**30**(7):485-491

[45] Rossi SJ, Neues LC, Kicokbusch TG. Thermodynamic and energetic evaluation of a heat pump applied to the drying of vegetables. In: Mujumdar AS, editor. Drying '92. Amsterdam: Elsevier Science; 1992. pp. 1475-1478

[46] Shi Q-L, Xue C-H, Zhao Y, Li Z-J, Wang X-Y. Drying characteristics of horse mackerel (*Trachurus japonicus*) dried in a heat pump dehumidifier. Journal of Food Engineering. 2008;**84**(1):12-20

[47] Soponronnarit S, Nathakaranakule A, Wetchacama S, Swasdisev IT, Rukprang P. Fruit drying using heat pump. RERIC International Energy Journal. 1998;**20**:39-53

[48] Teeboonma U, Tiansuwan J, Soponronnarit S. Optimization of heat pump fruit dryers. Journal of Food Engineering. 2003;**59**(4):369-377

[49] Vazquez G, Chenlo F, Moreira R, Cruz E. Grape drying in a pilot plane with heat pump. Drying Technology. 1997;**15**(3-4):889-920

[50] Zhang X, Mao Z, Sun L. Heat pump fluidized bed drying of agricultural materials. In: Annual Meeting of

American Society of Agricultural and Biological Engineers. 2004

[51] Aktaş M, Khanlari A, Aktekeli B, Amini A. Analysis of a new drying chamber for heat pump mint leaves dryer. International Journal of Hydrogen Energy. 2017;**42**(28):18034-18044

[52] Erbay Z, Hepbasli A. Application of conventional and advanced exergy analyses to evaluate the performance of a ground-source heat pump (GSHP) dryer used in food drying. Energy Conversion and Management. 2014;**78**:499-507

[53] Taşeri L, Aktas M, Sevik S, Gulcu M, Uysal Seckin G, Aktekeli B. Determination of drying kinetics and quality parameters of grape pomace dried with a heat pump dryer. Food Chemistry. 2018;**260**:152-159

[54] Omer AM. Ground-source heat pump system and applications. Renewable and Sustainable Energy Reviews. 2008;**12**:344-371

[55] Hancioglu KE, Hepbasli A. Exergetic evaluation of drying of laurel leaves in a vertical ground-source heat pump drying cabinet. International Journal of Energy Research. 2007;**31**:245-258

[56] Kumar M, Sansaniwal SK, Khatak P. Progress in solar dryers for drying various commodities. Renewable and Sustainable Energy Reviews. 2016;**55**:346-360

[57] Sarsavadia PN. Development of a solar-assisted dryer and evaluation of energy requirement for the drying of onion. Renewable Energy. 2007;**32**:2529-2547

[58] Juraev ET, Akhatov JS. Study of mixed-mode forced convection solar drier. Applied Solar Energy. 2009;**45**:61-64

[59] Slama RB, Combarnous M. Study of orange peels dryings kinetics and development of a solar dryer by forced convection. Solar Energy. 2011;**85**:570-578

[60] Singh PL. Silk cocoon drying in forced convection type solar dryer. Applied Energy. 2011;**88**:1720-1726

[61] Banout J, Ehl P, Havlik J, Lojka B, Polesny Z, Verner V. Design and performance evaluation of a double-pass solar drier for drying of red chilli (*Capsicum annuum* L.). Solar Energy. 2011;**85**:506-515

[62] Ferreira AG, Charbel ALT, Silva JG. Experimental analysis of a hybrid dryer. Thermal Engineering. 2007;**6**:3-7

[63] Lopez-Vidana EC, Mendez-Lagunas LL, Rodrıguez-Ramırez J. Efficiency of a hybrid solar-gas dryer. Solar Energy. 2013;**93**:23-31

[64] Reyes A, Mahn A, Vasquez F. Mushrooms dehydration in a hybrid-solar dryer, using a phase change material. Energy Conversion and Management. 2014;**83**:241-248

[65] Chaturvedi SK, Shen JY. Thermal performance of a direct expansion solar assisted heat pump. Solar Energy. 1984;**33**(2):155-162

[66] Morrison GL. Simulation of packaged solar heat-pump water heaters. Solar Energy. 1994;**53**(3):249-257

[67] Kuang YH, Wang RZ, Yu LQ. Experimental study on solar assisted heat pump system for heat supply. Energy Conversion and Management. 2003;**44**(7):1089-1098

[68] Troger JM, Butler JL. Drying peanuts with intermittent air flow. Transactions of ASAE. 1980;**23**(1):197-199

[69] Gan SH, Ng MX, Tham TC, Chua LS, Aziz R, Baba MR, et al. Drying characteristics of *Orthosiphon*

stamineus Benth by solar-assisted heat pump drying. Drying Technology. 2017;**35**(14):1755-1764

[70] Best R, Soto W, Pilatowsky I, Gutierrez LJ. Evaluation of a rice drying system using a solar assisted heat pump. Renewable Energy. 1994;**5**(1):465-468

[71] Şevik S, Aktaş M, Doğan H, Koçak S. Mushroom drying with solar assisted heat pump system. Energy Conversion and Management. 2013;**72**:171-178

[72] Mortezapour H, Ghobadian B, Minaei S, Khoshtaghaza MH. Saffron drying with a heat pump-assisted hybrid photovoltaic-thermal solar dryer. Drying Technology. 2012;**30**(6):560-566

Chapter 8

Bacteriocins of Lactic Acid Bacteria as Potent Antimicrobial Peptides against Food Pathogens

Parveen Kaur Sidhu and Kiran Nehra

Abstract

An ever-growing demand for food products with minimal chemical additives has generated a necessity for exploring new alternatives for food preservation. In this context, more recently, bacteriocins, the peptides having antimicrobial property, synthesized ribosomally by numerous bacteria have been attracting a lot of attention. They are known to possess the potential to restrict the growth of microorganisms causing food spoilage without causing any harm to the bacteria themselves owing to the presence of self-defensive proteins. In particular, the bacteriocins of lactic acid bacteria have been considered harmless and safe for consumption and are indicated to evade the development of unwanted bacteria. Use of bacteriocins as biopreservatives has been studied in various food industries, and they have been established to elevate the shelf life of minimally processed food items by exerting killing mechanism. They restrict the growth of undesirable bacteria by breaking the target cell membrane and finally resulting into pore formation. The current article provides an insight on bacteriocins of lactic acid bacteria, their biosynthesis, mechanism of action, and promising applications of these antimicrobial peptides in the food sector.

Keywords: antimicrobial peptides, bacteriocins, food-preservatives, food spoiling pathogens, lactic acid bacteria

1. Introduction

Most of the commercially available food preservatives are synthesized chemically, causing a lot of toxic effect on human health. The increase in the consumption of packed food which require preservatives for long-term storage has created a need for the production of minimally processed food free of any chemical additives. The present-day food industry is therefore looking for natural substitutes that have no adverse effect on consumers and on the surroundings to replace the existing chemical preservatives. Lactic acid bacteria (LAB) have been known since ages for their use in traditional fermented foodstuff owing to their capability to bring enviable changes in flavor, taste and for inhibiting food spoiling pathogenic microorganisms, thus making them an attractive natural biopreservative [1]. LAB are in general considered to be food grade microorganisms and assumed to be secure for human consumption as they get degraded when they come in contact with human gut by the action of proteases; hence, also contemplated as generally regarded as safe (GRAS) organisms. These are known to show preservative effect owing to their capability to produce hydrogen peroxide, organic acids, diacyls, and bacteriocins. Among all these, the antimicrobial compounds, bacteriocins, have received a greater attention as natural preservatives because nearly all of them are heat tolerant up to a particular temperature range and are amenable to proteolytic inactivation [1].

The bacteriocins are known to be produced by both positively and negatively Gram stained bacteria, however, the highest number of bacteriocins studied and identified are reported to be the antimicrobial peptides of Gram positive bacteria. A freely available database, BACTIBASE, which is totally dedicated to bacteriocins, consists of almost 177 sequences of bacteriocins out of which 156 belong to Gram positive bacterial strains, and the remaining 18 belong to Gram negative [2]. Production of bacteriocin depends on their genetic determinants; genes for bacteriocin production are reported to be localized either on plasmid or chromosomes [3]. These antibacterial peptides are synthesized on ribosomes and are known to show a bactericidal activity against identical or approximately related species [1]. Bacteriocins confer their bactericidal mode of action by creating a pore in the target cell, thus helping in combating the food pathogens.

2. Bacteriocins: antimicrobial peptide derivatives of lactic acid bacteria

During exponential growth, a huge number of both Gram positive and Gram negative bacteria produce peptides with antimicrobial activities. A universal term "bacteriocin" is used for these protein-like substances with bactericidal activity restricted to related species [4]. However, bacteriocins derived from Gram positive bacteria possess a moderately broader spectrum of antimicrobial activity. These proteinaceous compounds are degraded in the gastrointestinal tract of humans by protease digestion; hence, they become inactive in the human gut and do not pose any harmful effect. Production of bacteriocins is considered as an advantage by the food producers as they are efficient in killing or restricting pathogenic bacteria that struggle for the same habitat in a sufficient amount [5, 6].

Although several bacterial species produce bacteriocins, but LAB group of bacteria include some of the most important genera responsible for an elevated amount of bacteriocin production effective in the food industry [7]. LAB are considered as a heterogeneous cluster of Gram positive, non-aerobic but aero-tolerant, non-sporulating, catalase-negative, acid-tolerant, bacteria having low GC-content, which produce lactic acid by fermentation of sugars. They are recognized to be obtained from an array of different sources, for instance, meat, milk products, vegetables [6], grains, plants [8, 9] mucosal surface of animals, beverages, and from various fermented foods [10, 11]. LAB produce various metabolic products with antimicrobial properties, but current research is more focused on the production of bacteriocins which are toxic proteinaceous compounds and are known to exhibit a strong bactericidal activity against closely-related species; even small doses of bacteriocins are reported to be highly efficient in eradicating food spoiling bacteria [12–14].

2.1 Bacteriocins: classification

Bacteriocins produced from LAB were earlier categorized in four general classes on account of their characteristics by Klaenhammer et al. [12]. However, a more recent modified scheme of classification of LAB has been formulated by Alvarez-Sieiro et al. [13] which is now more acceptable (**Figure 1**). It includes only three major classes: Class I which includes lantibiotics, Class II which includes non-lantibiotics and class III which are basically large peptides (**Table 1**).

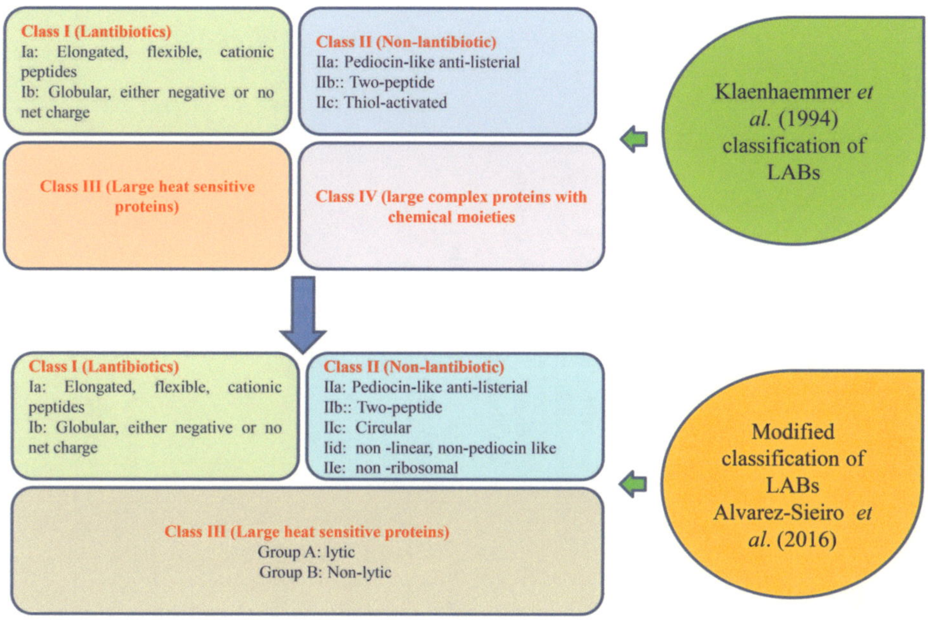

Figure 1.
Classification of bacteriocins produced by lactic acid bacteria.

Among all the three classes discussed in **Table 1**, the bacteriocins of class I and II have caught the attention of researchers; since these are more abundant and have been reported to have potential industrial applications including application in the food sector [15]. However, till date, only Nisin (class I) and Pedicon PA-1 (class II) are marketed commercially as a food additive. Nisin (marketed as Nispalin) is commercially approved to be used as a preservative in many food items in almost 50 countries in the world and Pediocin PA-1 (commercialized as Alto®2341) is authorized for its use in meat products [4].

2.1.1 Important commercialized bacteriocins: nisin and pediocin

Nisin is a 34 amino acid polypeptide having a molecular mass of 3354 Daltons. Synthesis of nisin is intricate, involving transcription, followed by post-translational modifications, secretion and final processing. It is synthesized as a pro-pre-peptide of 57 amino acid residues, where few residues located at terminal end function as the amino-terminal signal sequence of nisin molecule [1]. The other end of the nisin (carboxyl-terminal region of 34 residues) contains threonine, serine, and cysteine which are involved in the generation of reformed amino acids such as lanthionine, methyllanthionine, dehydroalanine and dehydrobutyrine, which are present in fully matured nisin [1] (**Figure 2a**). Nisin generally occurs in the form of a dimer (7 kDa), but may be present as a tetramer (14 kDa) at times. Nisin has two variants namely, nisin A and nisin Z. The difference between these two variants is because of a difference of one amino acid residue at the 27th position, histidine and asparagine being the 27th amino acid for nisin A and nisin Z, respectively. Nisin is predominantly used in dairy products and in canned food, and is particularly efficient when used in the production of cheese, restricting the growth of heat-tolerant sporulating strains [19]. It is also reported to be effective against mastitis-causing Gram positive organisms [20]. *Lactococci*, producer of nisin occur naturally in cheese and raw milk. Apart from

Class	Subclass	Characteristic features	Major representatives	References
I (Lantibiotics)	Small, heat-tolerant, and post-translationally modified peptides.		Nisin A, Enterocin W	[6, 15]
	Ia	Elongated, flexible, cationic peptides that generally perform their activity via pore formation and disruption of cell membrane.		
	Ib	Normally compact, with globular structures, either having negative charge or no net charge. These are enzyme inhibitors, immunologically active and act by interrupting the essential enzymatic reactions in the target species.		
II (Non-lantibiotics)	Heat-resistant antimicrobial peptides, devoid of any modified amino acid residues.			[13, 16, 17]
	IIa	Pediocin-like, active against Listeria spp., also known as listeria-active peptides.	Pediocin PA-1/Ach Sakacin A Carnobacteriocin X	
	IIb	It consists of complex of two different inactive peptides, activated by complex formation of these two inactive peptides.	Lactococcin G, Lactacin F Plantaricins EF and Plantaricin JK	
	IIc	Mostly circular peptides, N-terminal and C-terminal covalently linked to each other to form stable structure.	Acidocin B, Carnobacteriocin A, Enterocin P	
	IId	Non-pediocin, one-peptide, non-linear, sec-dependent bacteriocins.	Epidermicin NI01	
	IIe	Non-ribosomal having post-translational modifications at serine-rich C-terminal region.	Microcin E492	
III	Large in size, heat-sensitive, usually multi-domain with catalytic properties and have domain-type structure. Different domains in their domain structure function in different manner: i. binding to the receptor, ii. translocation, and iii. antimicrobial activity.			[13, 18]
	Group A	lytic enzymes that act by causing cell wall lysis.	Enterolysin A	
	Group B	Non-lytic proteins.	Caseicin 80	

Table 1.
Classification of bacteriocin produced by lactic acid bacteria.

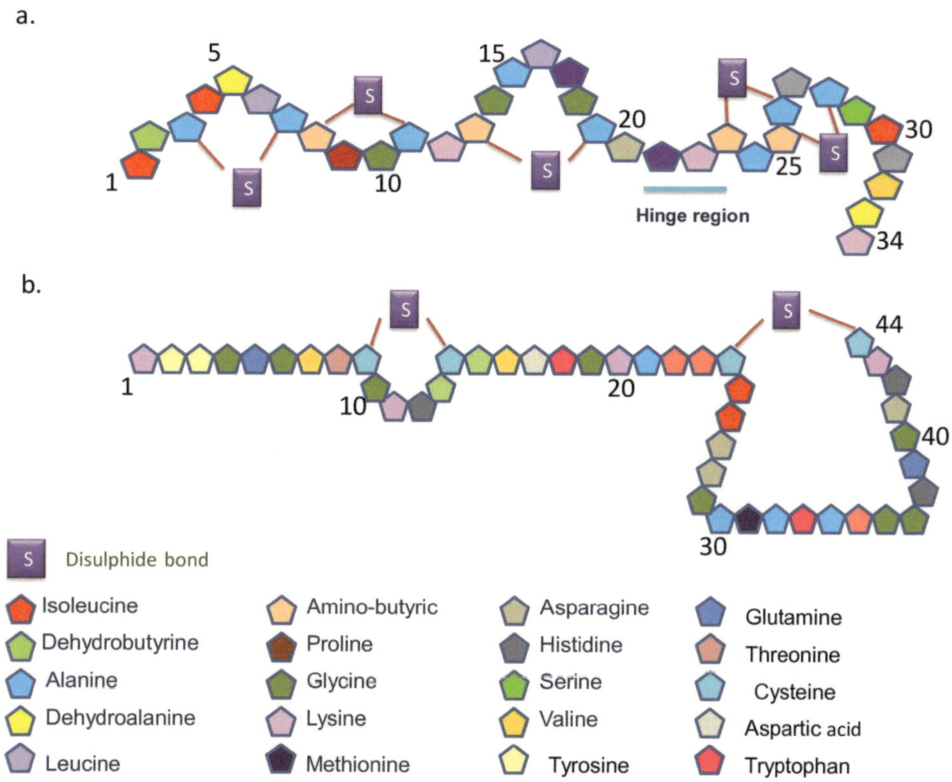

Figure 2.
Structure of (a) Nisin, showing the existence of post-translationally modified amino acids and (b) Pediocin PA-1.

nisin, other bacteriocins produced from genus *Lactococcus* are also of economic importance. Lacticin 3147 is an example of bacteriocin derived from *Lactococcus spp*. which works efficiently in a broad array of pH and comprises a wide antimicrobial action spectrum over Gram positive bacteria [5].

In contrast, Pediocin PA-1 is derived from the genus *Pediococcus* which is naturally found in many plant sources [8]. Strains of *Pediococcus* have been used since a long time as inoculum for the manufacture of fermented foods of natural sources [19]. Pediocin PA-1 shows an inhibitory response against few Gram positive bacteria but is reported to show strong antimicrobial activity against *Listeria monocytogenes* which is one of the chief food spoiling bacteria [4]. Pediocin PA-1 comprises of 44 amino acids with no post-translational modifications and has a molecular weight is 4646.95 Daltons (**Figure 2b**). The genes responsible for pediocin PA-1 biosynthesis are positioned on a 3.5 kb DNA segment of plasmid, and comprise four genes; (i) pedA (ii) pedB (iii) pedC and, (iv) pedD.

Although, nisin is approved by the Food and Drug Administration (FDA) for use as a food preservative, its use is limited to acidic foods only, because of its low solubility and constancy at high and neutral pH. This was the major reason for the search for bacteriocins from other species which led to the discovery of bacteriocins from *Pediococcus* species [21]. These bacteriocins are more effectual than nisin in inhibiting food spoiling pathogens, but their antibacterial spectrum is not as broad as nisin [22, 23]. Thus, although both nisin and pediocin play a significant role as a biological preservative in the food processing industry, their glaring limitations have caused the rise of research on bacteriocins from other classes as well.

3. Bacteriocins: genetics and biosynthesis

Bacteriocins produced from LAB are synthesized by genes encoded by either chromosomes or plasmids. For example, Plantaricin 423 [24] is plasmid encoded, while genes for enterocin A, sakacin P, divercin V41 and carnobacteriocin B2 and BM1 [25, 26] are localized on bacterial chromosomes. Plasmids concerned with the production of bacteriocins differ significantly in size. Some of them have been identified to carry the genetic determinants for numerous bacteriocins [3, 27].

3.1 Genetic constitution of bacteriocin operons

The genes accountable for synthesis of bacteriocin are often localized in single or multiple operons and are individually transcribed [28]. Most commonly, two types of genes are involved in production of bacteriocins: (i) structural genes and (ii) genes which encode for immunity protein. However, in some cases, apart from these two genes, other specific export machinery and regulatory genes are also required, thus making the bacteriocin operon much more intricate [29, 30].

3.1.1 Class I bacteriocin operon

Class I bacteriocin operon can be localized either on the bacterial chromosome or the plasmid. Few examples of lantibiotics with plasmid genetic determinants include: lacticin 481 [31] encoded by genes localized on 70 kb plasmid and the two-component lacticin 3147 [32] encoded on 63 kb plasmid, both of which are produced by Lactobacillus lactis. Class I bacteriocins are considered to be more complex than class II non-lantibiotics as they require supplementary enzyme encoding genes for post-translational modifications. The biosynthesis of lantiobiotics involves the translation of pre-peptide which undergoes few modifications, and then the modified pre-peptide moves onto the other side of the membrane, and the amino-terminal signal peptide is cleaved via proteolytic enzymes present in the cytoplasm. The best described lantibiotic is nisin, whose genetic determinants are reported to be localized on conjugative transposon Tn5276 contained by the bacterial chromosomes. Genes which aid in nisin production and immunity are transferred conjugally and it is reported to be situated in a nucleotide segment of size 8.5 kb. The gene cluster of nisin is designated as nisABTCIPRKFEG, and contains eleven genes which include structural genes, immunity protein encoding self-defensive genes, transporter genes and response regulator genes [3, 30]. Biosynthesis of nisin involves following steps: (a) firstly nisin A undergoes translation to form pre nisin A (b) The partially formed nisin A is then changed to form precursor nisin A via the proteins encoded by both nisB and nisC and (c) lastly, the precursor nisin is transported extracellularly by the nisT and nisP gene products by cleaving the leader peptide at the same time to obtain the end product, Nisin A (**Figure 3a**) [3, 33].

3.1.2 Class II bacteriocin operon

The genes encoding for structural proteins of class II bacteriocins are mostly localized on plasmids, for example, pediocin PA-1 and pediocin AcH [4] extracted from Pediococcus acidilactici strains. Bacteriocins belonging to class II usually require 2 to 8 genes for their production. The genes concerned with their production, secretion and immunity are: structural gene, immunity gene, ATP-binding cassette (ABC) protein encoding genes and its accessory proteins. Pediocin AcH, the bacteriocin representing class IIa is organized in a gene segment of 3500 bp

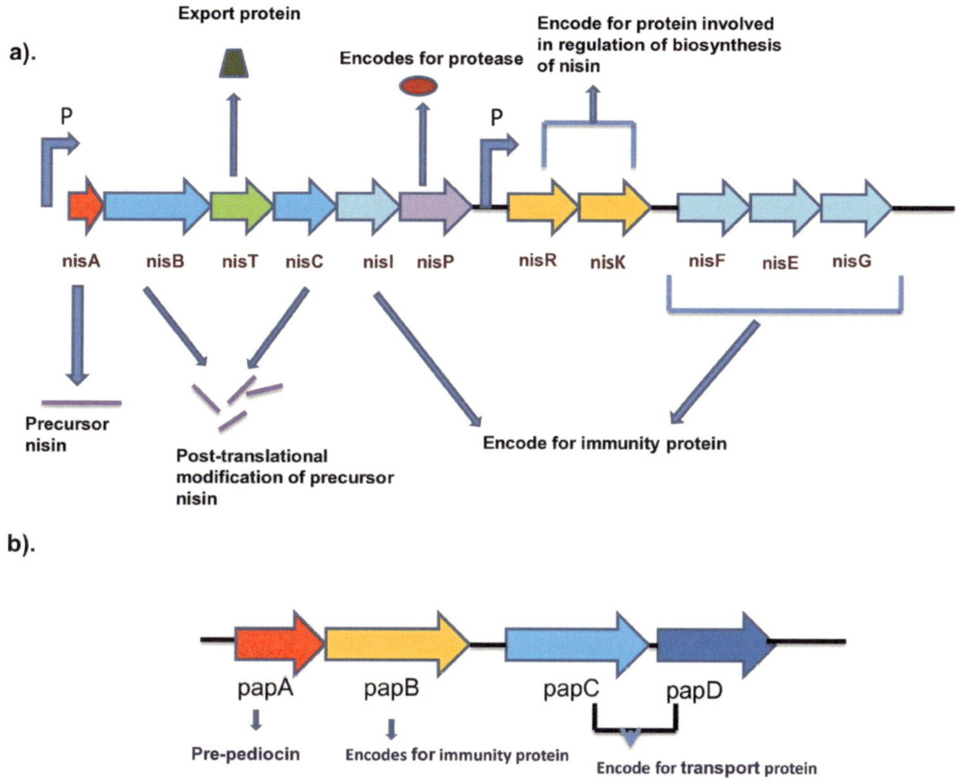

Figure 3.
Schemmatic structure of genes concerned with the synthesis of bacteriocins. (a) Nisin operon (b) Pediocin AcH operon. P: Promoter region.

having four genes designated as (i) papA, (ii) papB, (iii) papC and (iv) papD [3] (**Figure 3b**).

The genes involved in synthesis generally produce inactive pre-pro peptide having an amino-terminal leader peptide [29, 34]. The gene which imparts immunity is transcribed in parallel, along with the bacteriocin precursor. The ABC transporter and its accessory proteins which aid in the export of bacteriocins are encoded by a group of genes typically localized on operon present in closer proximity to immunity genes [3, 12, 29]. The leader peptide sequence of nearly all the bacteriocins of class IIa is of double glycine-type (GG-type), though, several of them have also been found to possess a sec-type amino-terminal signal sequence, for example, bacteriocin 31, listeriocin 743A, and enterocin P [35]. These bacteriocins are understood to be released out by sec-dependent exporting system. The ABC tranposters are assumed to aid in the recognition of GG-type leader peptides [29]. ABC transporters during transmembrane translocation consequently remove the GG-leader sequence and the fully active bacteriocin is afterwards secreted. Class IIa bacteriocin can be expressed heterologously through another secretion system by exchanging the leader peptide [29].

3.2 Biosynthesis of bacteriocins

Bacteriocins or antimicrobial peptides are primarily biosynthesized as pre-peptides which are biologically non-active, having amino-terminal signaling peptide which remains linked to carboxyl-terminal pro-peptide [13, 36]. This pre-pro-peptide is encoded by the structural genes. The pre-peptide is directed towards the maturation and transportation of the protein by the signal peptide which behaves as recognition spot and keeps the producing strain protected by being in an inactive state within the bacterial cells. Additionally, the signal peptide interacts with pro-peptide domain to make sure that it is in the proper state for further interface with modification machinery [24, 32]. The structural genes play an important role in bacteriocin biosynthesis and are generally succeeded by the immunity genes.

To defend the bacteriocin producing strains from the killing action of their self- bacteriocins, a self-defensive system known as immunity has been evolved by these bacteria. The protection from bacteriocins is gained via production of specific immunity proteins via immunity genes. These genes encoding immunity protein lie closer to the structural and accessory genes of bacteriocins [19]. Dissimilarity in expression and presence of these genes accounts for huge disparity in sensitivity demonstrated by LAB for bacteriocins. The size of immunity proteins lie in the range of 51 to 150 amino acids. An immunity gene encodes the immunity protein which lies downstream to the structural protein coding genes usually in the same operon. In the bacterial cell, transcription of both the bacteriocins and their immunity genes are regulated in parallel as defensive system can be stopped concurrently. Immunity of LAB depends on the definite immunity protein, Lan I [19], which basically remains bound to the exterior side of the plasma membrane. It grants protection to the producer strains by shutting off the membrane pores formed by the bacteriocin molecules and transferring them to the neighboring medium, hence, regulating the amount of bacteriocin present on the cell surface upto a critical level.

Bacteriocins are transported extracellularly with the aid of transporter proteins, which are basically from ABC transporter family. In prokaryotic and eukaryotic organisms, ABC transporters assist in the oozing of an ample range of products. These transporters include (i) bacterial importers, which aid in transportation of vitamins, oligopeptides, sugar, phosphate, amino acids and metallic ions, (ii) eukaryotic exporters, which enable the transportation of lipotropic drugs, pigments and peptides, and (iii) bacterial exporters, which carry large toxic protein, polysaccharides, heme molecules, and high and low molecular weight antibiotics [34]. The bacteriocin ABC transporters facilitate the exclusion of substrate from the signal peptide and its transportation across the cytoplasmic membrane hence, performing dual functions. This aids in effectively preventing the active and fully formed bacteriocins from enduring inside the cytoplasm.

The bacteriocin synthesis is regulated via multi-component regulatory system that involves (i) a signal peptide (ii) a cell surface bound receptor to which signal peptide binds and (iii) a response regulator. The signal peptide binds with histdine-kinase receptor and the binding leads to phosphorylation, signaling surge that finally targets the phosphate residue towards the response regulator that then induces the gene expression and production of bacteriocins by binding with the promoter region [34].

4. Bacteriocins: mode of action

Bacteriocins are recognized for restraining the growth of pathogenic microorganisms by forming pores in the cell envelope, and are known to be highly effective against Gram positive bacteria. Their activity against Gram negative bacteria is reported to be very less because of the outer wall of bacteria, which might possibly block the site of bacteriocin action [37]. Bacteriocins are proposed to exert their inhibitory effect using different mechanisms which include (i) a change in enzymatic activity (ii) restricting spore germination (iii) anionic carrier inactivation and (iv) with the aid of selective and non-selective pore formation in cell membrane [14, 37]. However, most of the bacteriocins follow the fourth model and act by targeting the cell envelope.

The primary interface connecting the bacteriocins and the target cell membrane involves either of the following two mechanisms: (a) attachment of bacteriocin to the membrane bound receptor molecule and/or (b) interface amidst the positively charged amino acids of bacteriocins and negatively charged phospholipid molecules on the cell membrane [29, 38]. In the first mechanism, the bacteriocins often need a receptor molecule on the surface of the target organism which varies among different species and sub-species [39]. The second mechanism involves three basic steps for the bactericidal action: binding, insertion and pore formation. In the binding and insertion step, when the bacteriocins come in contact with the target membrane, their C-terminal consisting of hydrophobic amino acids penetrates the hydrophobic region of the bacterial membrane and binds with the mannose phosphotransferase permease which ultimately leads to the leakage of the membrane. The interactions among phospholipids (negatively charged) in the target cell membrane and groups of amino acids (positively charged) in the bacteriocin are chiefly involved in the binding and insertion step. The bacteriocin finally creates pore in the target cell and allows the outflow of rather large molecules. Formation of pores eventually leads to the ionic imbalance, disturbing nucleic acid content and leakage of inorganic phosphates [19, 40]. The preliminary interruptions stimulate the dispersion of the proton motive force (PMF) that ultimately leads to the disturbance of pH and transmembrane potential of the cell. But this mode of action does not work for negatively stained bacteria as their bacterial membrane consists of lipopolysaccharide as an outer membrane which differs from Gram positive bacteria. Some researchers have reported that when bacteriocins are combined with compounds that have the capability to disrupt the outer membrane such as surfactants like Triton X-100 or EDTA, they can be rendered active against Gram negative bacteria [5].

Formation of pore in the cell membrane induced by the bacteriocins has been expected to take place by three models [38]: (i) a wedge-like model (ii) a barrel-stave like model and (iii) carpet model. A wedge-like model involves insertion of nisin and lipids via proton motive force. In barrel-stave like model, upon insertion in the membranes, bacteriocins arrange themselves to make a bunch of α helical peptides. The interior wall of the pore is formed by the inner hydrophilic faces of these peptides and the outer hydrophobic surface faces the membrane lipids. On the other hand, in carpet model, the pore is induced by peptides. In this model, an individual peptide might get assembled analogous to the cell surface and hinders the bilipid organization of the cell membrane and hence results in transient permeability due to strong phospholipid mobilizing activity [41]. For the induction of pore formation, class I bacteriocins are thought to act by means of wedge-like model, while in bacteriocins belonging to class II, pore formation may be either permitted via barrel-stave like model, or the carpet model [38]. Pore formation in class IIa bacteriocins is induced through barrel-stave like model probably because of the putative transmembrane helices in the peptide structure of these bacteriocin molecules (**Table 2**) [5, 29].

4.1 Bactericidal mode of class I bacteriocins

Bacteriocins belonging to class I are known to act by dual killing mechanisms, both of which have the same end results. Nisin is most researched antimicrobial peptide of this class. It exerts the bactericidal mode of action since it can diffuse easily through the anionic lipid membrane. It causes pore formation in the target membrane by coming in contact with the lipid II; a peptidoglycan precursor. The

Bacteriocin	Producing strain	Protein sequence	Killing mechanism
Class I bacteriocins			
Nisin	*Lactococcus lactis subsp. Lactis*	(40 amino acids) ITSISLCTPG CKTGALMGCN MKTATCHCSI HVSK	Dual killing mechanism leading to cell death.
Lacticin 481	*Lactococcus lactis*	(30 amino acids) KGGSGVIHTI SHECNMNSWQ FVFTCCS	Inhibits peptidoglycan formation.
Lacticin 3147	*Lactococcus lactis* subsp. *lactis*	(30 amino acids) TTPATPAISI LSAYISTNTC PTTKCTRAC	Interaction with the cell membrane leads to the pore formation even at minimal concentrations.
Lactocin-S	Lactobacillus sakei L45	(40 amino acids) STPVLASVAV SMELLPTASV LYSDVAGCFK YSAKHHC	Form transmembrane pores in the target bacterial cytoplasmic membrane.
Plantaricin A	*Lactobacillus plantarum*	(30 amino acids) AYSLQMGATA IKQVKKLFKK WGW	Pore formation by barrel stave model.
Acidocin J1132 β	*Lactobacillus acidophilus*	(30 amino acids) GNPKVAHCAS QIGRSTAWGA VSGA	Form cell membrane channel by barrel stave mechanism.
Class II bacteriocins			
Plantaricin J	*Lactobacillus plantarum*	(30 amino acids) GAWKNFWSSL RKGFYDGEAG RAIRR	Membrane leakage and cell death.
Acidocin 8912	*Lactobacillus acidophilus*	(30 amino acids) KTHYPTNAWK SLWKGFWESL RYTDGF	Causes scattering of proton motive force and by pore formation.
Bavaricin-A	*Lactobacillus sakei*	(50 amino acids) KYYGNGVHXG KHSXTVDWGT AIGNIGNNAA ANXATGXNAG G	Show killing action and inhibit the growth of *Listeria monocytogenes*
Pediocin PA-1	*Pediococcus acidilactici*	(50 amino acids) KYYGNGVTCG KHSCSVDWGK ATTCIINNGA MAWATGGHQG NHKC	Pore formation (effective inhibitor of *Listeria monocytogenes*)

Bacteriocin	Producing strain	Protein sequence	Killing mechanism
Enterolisin A	*Enterococcus faecalis*	(340 amino acids) MKNILLSILG VLSIVVSLAF SSYSVNAASN EWSWPLGKPY AGRYEEGQQF GNTAFNRGGT YFHDGFDFGS AIYGNGSVYA VHDGKILYAG WDPVGGGSLG AFIVLQAGNT NVIYQEFSRN VGDIKVSTGQ TVKKGQLIGK FTSSHLHLGM TKKEWRSAHS SWNKDDGTWF NPIPILQGGS TPTPPNPGPK NFTTNVRYGL RVLGGSWLPE VTNFNNTNDG FAGYPNRQHD MLYIKVDKGQ MKYRVHTAQS GWLPWVSKGD KSDTVNGAAG MPGQAIDGVQ LNYITPKGEK LSQAYYRSQT TKRSGWLKVS ADNGSIPGLD SYAGIFGEPL DRLQIGISQS NPF	Cell wall degrading enzyme shows bacteriolytic mode of action.
Class III bacteriocins			
Helveticin J	*Lactobacillus helveticus*	(330 amino acids) MKHLNETTNV RILSQFDMDT GYQAVVQKGN VGSKYVYGLQ LRKGATTILR GYRGSKINNP ILELSGQAGG HTQTWEFAGD RKDINGEERA GQWFIGVKPS KIEGSKIIWA KQIARVDLRN QMGPHYSNTD FPRLSYLNRA GSNPFAGNKM THAEAAVSPD YTKFLIATVE NNCIGHFTIY NLDTINEKLD EKGNSEDVNL ETVKYEDSFI IDNLYGDDNN SIVNSIQGYD LDNDGNIYIS SQKAPDFDGS YYAHHKQIVK IPYYARSKES EDQWRAVNLS EFGGLDIPGK HSEVESIQII GENHCYLTVA YHSKNKAGEN KTTLNEIYEL SWN	Generally show bactericidal mode of action.

Table 2.
General overview of few important bacteriocins belonging to Lactic acid bacteria, their protein sequence and killing mechanism (Data obtained from BACTIBASE database of bacteriocins [http://bactibase.hammamilab.org]).

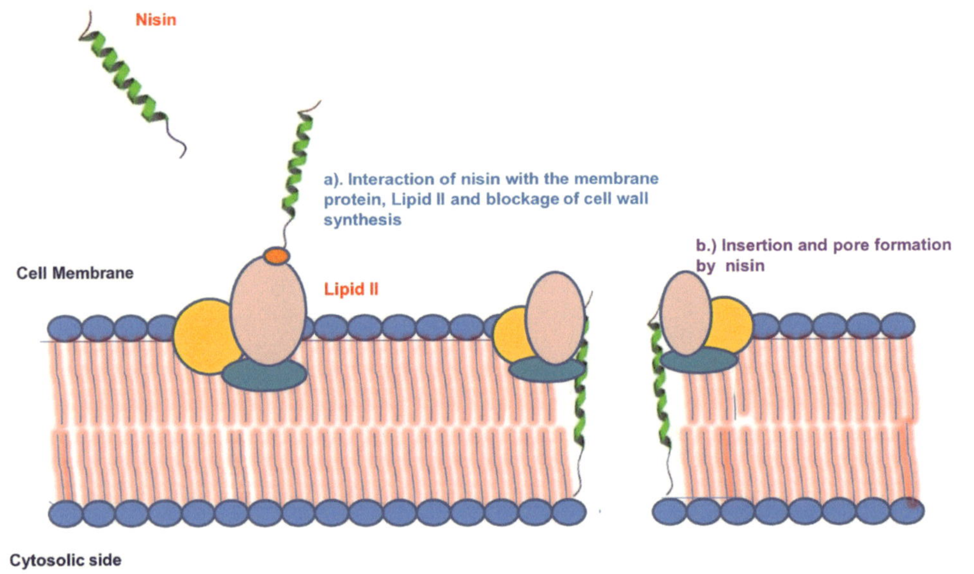

Figure 4.
Two-way killing mechanism of class I bacteriocins. (a) Nisin forms complex with the lipid II molecule and halts the synthesis of cell wall. (b) In second mechanism, nisin forms complex with lipid II molecules and initiates formation of pore, ultimately leading to cell death.

existence of lipid II increases nisin activity thereby causing membrane depolarization, disruption of bilayer organization and outflow of limited metabolites like, ions, nucleotides and amino acids, hence halting all the biosynthetic pathways and causing cell death [41]. Nisin is very effective against its target strain and has been reported to show antimicrobial activity even at nanomolar concentrations. When present in lower concentration, nisin gets attached to lipid II molecule and halts cell wall formation by enzyme activity which causes cell death. Furthermore, when nisin is present in higher amount, the nisin- lipid II complex causes sudden cell death by inducing pores in the cell membrane of bacteria (**Figure 4**). Hence, this complex assists in two modes of killing, involving both blockage of cell wall synthesis, and formation of pores (**Figure 4**).

4.2 Bactericidal mechanism of class II bacteriocins

Bacteriocins of class II (non-lantibiotics) show a killing action against related organisms and cause depolarization and cell death by easy insertion into the cell membrane due to their amphiphilic nature. These bacteriocins are capable of disturbing the pH gradient and transmembrane potential of cells which results in scattering of proton motive force; it is reported that this dispersal of proton motive force is believed to be their key action for exerting the fatal activity [42, 43]. The killing mechanism of class II bacteriocins varies among subclasses (**Table 3**) [39]. Class IId and class IIe mechanisms are still under study and are poorly understood till now.

5. Potential applications of bacteriocins in food sector

Bacteriocins produced by LAB have promising applications in the health, pharmaceutical and food sector. However, this section basically reviews the applications of bacteriocins in food safety. Currently, in food industries, bacteriocins are

DOI: http://dx.doi.org/10.5772/intechopen.95747

S.No.	Subclass	Killing mechanism	References
1.	IIa	i. These bacteriocins are highly potent against *L.monocytogenes.* ii. They act by binding to the mannose phosphotransferase system proteins which are present in the non-polar core of the cell membrane. iii. The conserved amino-terminus and carboxyl-terminus region is accountable for the inhibiotory activity against *L.monocytogenes* and few other desired organisms.	[28, 42]
2.	IIb	i. They are found to exert their killing effect against target organisms only when both the peptides show synergistic activity. ii. Killing mechanism of this subclass involves permeabilization of the target membrane which causes leakage of small cytoplasmic molecules. iii. They create comparatively small and specific pores than bacteriocins of class IIa.	[44]
3.	IIc	i. They show wide antimicrobial spectrum against food pathogens. ii. They act similar to the other bacteriocins in case of mode of action towards target organism, i.e, by causing membrane permeabilization, efflux of ions and leading to cell death.	[45]

Table 3.
Tabulated overview of Killing mechanism of subclass of class II bacteriocins.

extensively used as biological preservatives. Both Gram positive and Gram negative bacteria are efficient in producing the bacteriocins, but those originating from LAB have a greater importance in the food sector [46]. Bacteriocins produced by LAB apart from being regarded as safe also have QPS (qualified presumption of safety) status, and show deleterious effects against food pathogens even in nanomolar range [47, 48].

Bacteriocins are known to have potential preservative properties either when used alone or when combined with other preservation methods in the form of hurdle technology [4, 30, 36]. Bacteriocins as natural preservatives offer following advantages: (a) extended shelf life of food, (b) protection from economic loss due to food spoilage, (c) preservation of the nutrients and vitamins of food and thus maintenance of the flavor and taste of food (d) satisfaction of consumer demands and (e) stability at variable temperatures and heat. Although, the use of purified form of bacteriocin is the usually applied approach, however, the direct addition of LAB has also been found to be effective [36]. The bacteriocins can be applied by at least three different ways to advance the safety and quality of food, (i) by addition of purified or partially purified preparation of bacteriocins in food items, (ii) by addition of the product formerly fermented with a producer strain [19].

One of the essential applications of bacteriocin is bioactive packaging which protects the food from exterior contaminants. Bioactive packaging generally involves the assimilation of producer strains into the packaging substance that is expected to directly interact with food, thereby, helping to improve the storage life and food safety by restricting the growth of food spoiling organisms, mainly in meat and cheese [49]. The interaction of food surface with the packaging film allows easy diffusion of bacteriocins into the food making this method advantageous over drop-wise addition and sprinkling of bacteriocins on food material; as in the latter cases, the food components may hinder the antibacterial activity of bacteriocins [50–53]. Antimicrobial packaging films are able to be developed by straight-away inclusion of the antimicrobial peptide in the packaging of foodstuff, or by an addition of a packet having bacteriocin in the ready-to-eat packed food, which would be later released in the food product during storage period. Research on nisin coated packaging is increasingly being encouraged since the past few years. Neetoo et al. [54] studied the use of nisin-coated synthetic films on vacuum-packed cold-smoked salmon, and observed that the coating when carried out at a specific storage temperature resulted into a remarkable decrement in the survival rate of *L. monocytogenes*. This method is considered to be better than other methods of bacteriocin application as discussed before the bacteriocins can be deactivated, lose its activity by coming directly in contact with food components [55].

Bacteriocins of LAB are normally found to be potent inhibitors of pathogenic Gram positive bacteria, but when nisin and few other bacteriocins are treated with surfactants, they show wide activity against Gram negative bacteria as well. Nisin is the most researched bacteriocin and has been approved for its commercial production for use in cheese, canned vegetables, egg products etc. [56]. It is often used in acidic foods and also has been used in inhibiting undesirable microorganisms in beer and wine [1]. Nisin is also reported to be conjugated with few other preservative procedures such as heat or other types of bacteriocins. This process/technique is found to be more effective in eradicating food spoiling bacteria due to the increased antimicrobial spectrum. Nisin is preferably applied in aqueous form, as the powder form may have improper dispersal issues. Sea food industry is also being benefited from nisin in restricting the expansion of food spoiling microorganisms. A research by Bakkal et al. [57] showed the delayed growth of *L. monocytogenes* in cold smoked salmon when treated with nisin. A similar study carried out by Pei et al. [58] discussed the inhibitory action by nisin on the spoilage bacteria present in tangerine wine. Nisin in amalgamation with lactases has been found to be highly competent to restrict the flourishment of *L. monocytogenes*, this may be due to synergistic activity [59]. Use of nisin for non-acidic food and dairy products is very limited; therefore the need arises to identify new bacteriocins having ability to retain stability in different food systems and is also thermally stable.

Comparatively, only a few studies have been carried out on the applications of pediocin in the food industry. Pediocin PA- 1/AcH which is accepted for commercial use in meat products has been found to efficiently inhibit the development of *Listeria* species present in ice cream mixture, ground beef and sausage mix [60]. The addition of pediocins as preservatives in the food system aids in efficiently inhibiting the pathogenic food bacteria thereby guaranteeing the extended storage life of food, and safety of consumers [1, 4]. However, only a few bacteriocins have been approved at commercial level. The reason might be that, the newly identified antimicrobial peptides yet remain to be fully characterized, and further studies need to be carried out to gain a complete insight in their molecular mechanism implicated in bacteriocin production.

Bacteriocins are effective in inhibiting several other food spoiling pathogenic bacteria also. Chang and Chang [61] reported the restriction of growth of *Staphylococcus aureus* and *Escherichia coli* by Kimchi made with *Leuconostoc citreum* GJ7, a bacteriocin producing strain. Starter culture of LAB was also found to restrict the *Bacillus cereus* growth in rice fermentation [62] and growth of *Escherichia coli* and *Clostridium perfringens* were also reported to be inhibited by *Lactobacillus plantarum* and *Lactobacillus salivarius* on chicken feed media [63]. These are known to be the major trouble-causers to the food preservation industry. Although the use of bacteriocins has now started becoming popular in the food industry but their full potential has not yet been realized. They still suffer from several limitations as narrow antimicrobial spectrum, high dosage necessity, high production expenditure and low yield. To combat these limitations, research is now-a-days heading in the direction towards combining two or more classes of bacteriocins that aids in checking their efficacy as an improved/better preservative. Additionally, bacteriocins are also being conjugated with nanoparticles to increase their antimicrobial spectrum and for efficient delivery in target cells [30, 40, 64, 65]. These advancements are bound to contribute more towards new inventions and applications in food sector.

6. Future perspective

Bacteriocins have proven themselves to be potent antimicrobials produced by bacteria. Their applications in various sectors encourage for carrying out research for investigating different applications of bacteriocins in diverse areas. Currently, there is a critical need to combat the limitations of bacteriocins, most significantly, the narrow antimicrobial spectrum. For their effective use against various Gram-negative and food-borne pathogens, the already existing techniques need to be supplemented with new methods. There is a need to develop cost-effective methods for purification of bacteriocins and enhancing their production.

Studies conducted using nanotechnology techniques for enhancing antimicrobial spectrum and lowering the dosage requirement of bacteriocins are also encouraged. After that, evaluating the safety of nanoconjugates using various toxicity tests both *in-vivo* and *in-vitro* should be conducted to authorize their use in both food and pharmaceutical sectors. Furthermore, studying the interaction between bacteriocins and nanoparticles can promote the use of bacteriocins in future. Moreover, there is a need to introduce and design new open-access bacteriocin database, having information about newly discovered bacteriocins; their properties structure, and applications.

This will help the researchers to study already characterized bacteriocins in a simpler way. Additionally use of newly developed techniques viz., mass spectroscopy, bioinformatics studies will aid in understanding and characterizing the bacteriocins in more efficient ways.

7. Conclusion

Bacteriocins produced by LAB act as a weapon used by the bacterial cells to fight against other competing bacteria. This property of bacteriocin-producing LAB strains has been exploited by the scientists to combat several issues related to food spoilage and safety in the food sector. The current review is a concerted effort towards exploring in detail the biosynthesis, mechanism of action and existing applications of class I and class II bacteriocins produced by different LAB. However, although several studies describe the emerging and flourishing applications of bacteriocins in the health and food sector, but still, several hurdles limit the realization of their full potential in the commercial sector. Hence, the present research in the area of bacteriocins is getting more targeted towards looking for extraction and purification of more promising novel bacteriocins, and enhancement of antimicrobial spectrum of already existing purified bacteriocins. As an effort in this direction, although several bacteriocins have been studied as well as purified in the last few years, there is still a need as well as huge scope for continuing more dedicated research. A continued effort will surely enable the research fraternity to study and discover novel antimicrobial peptides with more promising properties.

Acknowledgements

We wish to express our sincere gratitude to Deenbandhu Chhotu Ram University of Science and Technology for providing Research facility.

Conflict of interest

The authors declare no conflict of interest.

Author details

Parveen Kaur Sidhu and Kiran Nehra*
Department of Biotechnology, Deenbandhu Chhotu Ram University of Science and Technology, Haryana, India

*Address all correspondence to: nehrakiran@gmail.com

References

[1] Jeevaratnam K, Jamuna M, Bawa AS. Biological preservation of food-Bacteriocins of lactic acid bacteria. Ind J Biotechnol 2005;4:446-454.

[2] Lagha AB, Haas B, Gottschalk M, Grenier D.Antimicrobial potential of bacteriocins in poultry and swine production. Vet res 2017;48:22-32.

[3] Dimov S, Ivanova P, Harizanova N. Genetics of Bacteriocins Biosynthesis by Lactic Acid Bacteria. Biotechnol Biotec Eq 2005;19:4-10.

[4] Silva CCG, Silva SPM, Riberio SC. Application of bacteriocins and protective cultures in dairy food preservation. Front Microbiol 2018;9:594. doi: 10.3389/fmicb.2018.00594

[5] Parada JL, Caron CR, Medeiros ABP, Soccol CR. Bacteriocins from Lactic Acid Bacteria: Purification, Properties and use as Biopreservatives. Braz arch biol Technol 2007; 50:521-542.

[6] Joshi VK, Sharma S, Rana NS. Bacteriocin from Lactic Acid Fermented Vegetables. Food Technol Biotechnol 2006;44:435-439.

[7] Daba GM, Elkhateeb WA. Bacteriocins of lactic acid bacteria as biotechnological tools in food and pharmaceuticals: Current applications and future prospects, Biocatal Agric Biotechnol 2020; 28:101750.

[8] Ray B. Pediococcus in fermented foods, in Food biotechnology microorganisms, edited by Y. H. Hui & G. G. Khachatourians, p. 745-796. New York: VCH Publishers, Inc; 1995.

[9] Alemayehu D, Hannon JA, Mc Auliffe O, Ross RP. Characterization of plant-derived lactococci on the basis of their volatile compounds profile when grown in milk. Int J Food Microbiol 2014;172:57-61.

[10] Kivanc M, Yilmaz M, Cakir E. Isolation and identification of lactic acid bacteria from boza, and their microbial activity against several reporter strains. Turk J Biol 2011;35: 313-324. doi:10.3906/biy-0906-67.

[11] Abegaz K. Isolation, characterization and identification of lactic acid bacteria involved in traditional fermentation of borde, an Ethiopian cereal beverage. Int J Food Nutr Sci 2014;1:7-15.

[12] Klaenhammer TR, Fremaux C, Hechard Y. Activité antimicrobienne des bactéries lactiques. In Bactéries Latiques, H. De Roissart and F. M. Luquet. France: Lorica; 1994.

[13] Alvarez-Sieiro P, Montalbán-López M, Mu D, Kuipers OP. Bacteriocins of lactic acid bacteria: extending the family. Appl Microbiol Biotechnol 2016;100:2939-2951. doi: 10.1007/s00253-016-7343-9

[14] Meade E, Slattery MA, Garvey M. Bacteriocins, Potent Antimicrobial Peptides and the Fight against Multi Drug Resistant Species: Resistance Is Futile?. Antibiotics (Basel). 2020;9(1):32.

[15] Yang S, Lin C, Sung CT, Fang J. Antibacterial activities of bacteriocins : application in foods and pharmaceuticals. Front microbial 2014;5:1-10.

[16] Cotter PD, Ross RP, Hill C. Bacteriocins—a viable alternative to antibiotics? Nat Rev Microbiol 2013;11:95-105. doi: 10.1038/nrmicro2937

[17] Sandiford S, Upton M. Identification, characterization, and

recombinant expression of epidermicin NI01, a novel unmodified bacteriocin produced by *Staphylococcus epidermidis* that displays potent activity against Staphylococci. Antimicrob Agents and Chemother 2012;56:1539-1547. doi: 10.1128/AAC.05397-11.

[18] Nilsen T, Nes IF, Holo H. Enterolysin A, a cell wall-degrading bacteriocin from *Enterococcus faecalis* LMG 2333. Appl Environ Microbiol 2003;69:2975-2984. doi: 10.1128/AEM.69.5.2975-2984.2003.

[19] Deegan LH, Cotter PD, Hill C, Ross P. Bacteriocins: Biological tools for biopreservation and shelf-life extension. Int Dairy J 2006;16: 1058-1071.

[20] Broadbent JR, Chou C, Guillies K., Kondo JK. Nisin inhibits several gram-positive, mastitis-causing pathogens. J Dairy Sci 1989; 72:3342-3345.

[21] Shin JM, Gwak JW, Kamarajan P, Fenno JC, Rickard AH, Kapila YL. Biomedical applications of nisin. J Appl Microbiol 2016; 120:14491465.

[22] Ray B, Miller KW. Pediocin: Natural food antimicrobial systems, p. 525-566. Boca Raton, FL: CRC Press; 2000.

[23] Da Costa RJ, Voloski FLS, Mondadori RG, Duval EH, Fiorentini AM. Preservation of Meat Products with Bacteriocins Produced by Lactic Acid Bacteria Isolated from Meat. J Food Qual 2019;1-12.

[24] Van Belkum MJ, Worobo RW, Stiles ME. Double-glycine-type leader peptides direct secretion of bacteriocins by ABC transporters: colicin V secretion in Lactococcus lactis. Mol Microbiol 1997;23:1293-1301.

[25] Quadri LEN, Kleerebezem M, Kuipers OP, de Vos WM, Roy KL, Vederas JC, Stiles ME. Characterization of a locus from Carnobacterium piscicola LV17B involved in bacteriocin production and immunity: evidence for global inducer-mediated transcriptional regulation. J Bacteriol 1997;179:6163-6171.

[26] Metivier A, Pilet MF, Dousset X, Sorokine O, Anglade P, Zagorec M, Piard JC, Marion D, Cenatiempo Y, Fremaux C. Divercin V41, a new bacteriocin with two disulphide bonds produced by Carnobacterium divergens V41: primary structure and genomic organization. Microbiol 1998;144:2837-44.

[27] Perez RH, Zendo T, Sonomoto K. Circular and Leaderless Bacteriocins: Biosynthesis, Mode of Action, Applications, and Prospects. Front Microbiol 2018; 9:2085.

[28] Fimland G, Johnsen L, Dalhus B Nissen-Meyer J. Pediocin-like antimicrobial peptides (class IIa bacteriocins) and their immunity proteins: biosynthesis, structure, and mode of action. J Pept Sci 2005;11: 688-696.

[29] Ennahar S, Sashihara T, Sonomoto K, Ishizaki A. Class IIa bacteriocins: biosynthesis, structure and activity. FEMS Microbiol Rev 2000;24: 85-106.

[30] Sidhu PK, Nehra K. Bacteriocin-nanoconjugates as emerging compounds for enhancing antimicrobial activity of bacteriocins. JKSUS 2019; 31:758-767.

[31] Rincé A, Dufour A, Uguen P, Le Penec JP, Haras D. Characterization of the lacticin 481 operon: the Lactococcus lactis genes lctF, lctE, and lctG encode a putative ABC transporter involved in bacteriocin immunity. Appl Environ Microbiol 1997;63(11):4252-4260.

[32] McAuliffe O, Ross RP, Hill C. Lantibiotics: structure, biosynthesis and

mode of action. FEMS Microbiol Rev 2001;25:285-308.

[33] Guinane C, Piper C, Draper LA, O'Connor PM, Hill C, Ross RP, Cotter PD. Impact of environmental factors on bacteriocin promotor activity in gut-derived *Lactobacillus salivarius*. Appl Environ Microbiol 2015; 81:7851-7859.

[34] Nes IF, Diep DB, Havarstein LS, Brurberg MB, Eijsink V, Holo H. Biosynthesis of bacteriocins in lactic acid bacteria. Antonie van Leeuwenhoek 1996; 70:113-128.

[35] Kalmokoff ML, Banerjee SK, Cyr T, Hefford MA, Gleeson T. Identification of a new plasmid-encoded sec-dependent bacteriocin produced by Listeria innocua 743. Appl Environ Microbiol 2001;67(9):4041-4047.

[36] Chen H, Hoover DG. Bacteriocins and their food applications. Compr Rev Food Sci F 2003;2:82-100.

[37] Martinez RCR, De Martinis ECP. Effect of Leuconosoc mesenteroides bacteriocin in the multiplication control of *Listeria monocytogenes*. Ciên Tecnol Aliment 2006;26:52-55.

[38] Moll GN, Konings WN, Driessen AJM. Bacteriocins: mechanism of membrane insertion and pore formation. Antonie van Leeuwenhoek 1999;76:185-198.

[39] Perez RH, Perez MTM, Elegado FB. Bacteriocins from lactic acid bacteria: A review of biosynthesis, mode of action, fermentative production, uses and prospects. Int J Phil Sci Technol 2015;8:61-67.

[40] Todorov SD, Powell JE, Meincken M, Witthuhn RC, Dicks LMT. Factors affecting the adsorption of Lactobacillus plantarum bacteriocin bacST8KF to *Enterococcus faecalis* and Listeria innocua. Int J Dairy Technol 2007;60:221-227.

[41] Bechinger B. Structure and functions of channel-forming peptides: magainins, cecropins, melittin and alamethicin. J Memb Biol 1997;156:197-211. Rodriguez E, Martinez MI, Horn N, Dodd HM. Heterologous production of bacteriocins by Lactic Acid Bacteria. Int J Food Microbiol 2003;80:101-116.

[42] Todorov SD. Bacteriocins from Lactobacillus plantarum–production, genetic organization and mode of action. A review. Braz J Microbiol 2009;40:209-221.

[43] Dicks L, Dreyer L, Smith C, van Staden AD. A Review: The Fate of Bacteriocins in the Human Gastro-Intestinal Tract: Do They Cross the Gut-Blood Barrier? Front Microbiol 2018; 9:2297. doi: 10.3389/fmicb.2018.02297.

[44] Oppegard C, Rogne P, Emanuelsen L, Kristiansen PE, Fimland G, Nissen-Meyer J. The two-peptide class II bacteriocins: structure, production, and mode of action. J Mol Microbiol Biotechnol 2007;13:210-219.

[45] Gabrielsen C, Brede DA, Nes IF, Diep DB. Circular bacteriocins: biosynthesis and mode of action. Appl Environ Microbiol 2014;80:6854-6862.

[46] Chikindas ML, Weeks R, Drider D, Chistyakov VA, Dicks LM. Functions and emerging applications of bacteriocins. Curr Opin Biotechnol 2018;49:23-28.

[47] Field D, Ross RP, Hill C. Developing bacteriocins of lactic acid bacteria into next generation biopreservatives. Curr Opin Food Sci 2018; 20:1-6.

[48] Kaya HI, Ozel B,̧ Sim̧sek O. A natural way of food preservation: Bacteriocins and their applications. InHealth and Safety Aspects of

Food Processing Technologies; Malik A, Erginkaya Z, Erten H, Eds.; Springer:Cham, Berlin, Germany, 2019.

[49] Schöbitz RP, Bórquez PA, Costa ME, Ciampi LR, Brito CS. Bacteriocin like substance production by Carnobacterium piscicola in a continuous system with three culture brooths. Study of antagonism against Listeria monocytogees in vacuum packaged salmon. Braz J Microbiol 2006;37:52-57.

[50] Feng X, Ng VK, Miks-Krajnik, M, Yang H. Effects of fish gelatin and tea polyphenol coating on the spoilage and degradation of myofibril in fish fillet during cold storage . Food Bioprocess Technol 2017; 10:89-102 .

[51] Misra NN, Yepez X, Xu L, Keener K. In - package cold plasma technologies . J Food Eng. 2019; 244:21-31.

[52] Moutiq R, Misra NN, Mendonca A, Keener K. In-package decontamination of chicken breast using cold plasma technology: microbial, quality and storage studies. Meat Sci 2020; 159:107942.

[53] Olatunde OO, Benjakul S, Vongkamjan, K. Shelf - life of refrigerated Asian sea bass slices treated with cold plasma as affected by gas composition in packaging . Int J Food Microbiol 2020; 324:1-11.

[54] Neetoo H, Ye M, Chen H. Potential antimicrobials to control *Listeria monocytogenes* in vacuum-packaged cold-smoked salmon pâté and fillets. Int J Food Microbiol 2008;123:220-227.

[55] Pilevar Z, Hosseini H, Beikzadeh S, Khanniri E, Alizadeh AM. 2020 . Application of bacteriocins in meat and meat products: an update. Curr Nutr Food Sci 2020; 16 120-133.

[56] Deshmukh PV, Thorat PR. Bacteriocins: A new trend in antimicrobial food packaging. Int J Adv Res Eng Appl Sci 2013; 2:1-12.

[57] Bakkal S, Robinson SM, Riley MA. Bacteriocins of Aquatic Microorganisms and their Potential Applications in the Seafood Industry. In Health and Environment in Aquaculture, p. 303-328. Croatia: InTech; 2012.

[58] Pei J, Jiang L, Dai H, Chen P. Application of Nisin – the Well-known Lactic Acid Bacteria Bacteriocin – against Spoilage Bacteria in Tangerine Wine. Czech J Food Sci 2016;34:488-494.

[59] Gálvez, A, López RL, Abriouel H, Valdivia E, Omar NB. Application of bacteriocins in the control of foodborne pathogenic and spoilage bacteria. Crit Rev Biotechnol 2008; 28:125-152.

[60] Motlagh A, Bukhtiyarova M, Ray B. Complete nucleotide sequences of pSMB74, a plasmid encoding production of pediocin AcH in Pediococcus acidilactici. Lett Appl Microbiol 1994; 18:305-312.

[61] Chang JY, Chang HC. Growth inhibition of food-borne pathogens by kimchi prepared with bacteriocin-producing starter culture. J Food Sci 2011; 76:M72-78.

[62] Yang Y, Tao WY, Liu YJ. Zhu, F. Inhibition of *Bacillus cereus* by lactic acid bacteria starter cultures in rice fermentation. Food Control 2008; 19:159-161.

[63] Murry AC, Hinton A, Morrison H. Inhibition of growth of *Escherichia coli, Salmonella typhimurium*, and *Clostridia perfringens* on chicken feed media by *Lactobacillus salivarius* and *Lactobacillus plantarum*. Int J Poult Sci 2004; 3:603-607.

[64] Sidhu PK, Nehra K. Bacteriocin-capped silver nanoparticles for enhanced antimicrobial efficacy against food pathogens. IET Nanobiotechnol, 2020a; 14: 245-252.

[65] Sidhu PK, Nehra K. Purification and characterization of bacteriocin Bac23 extracted from *Lactobacillus plantarum* PKLP5 and its interaction with silver nanoparticles for enhanced antimicrobial spectrum against food-borne pathogens. LWT- Food Sci Technol 2020; 110546.

Chapter 9

Clinical Approaches of Biomimetic: An Emerging Next Generation Technology

Kirti Rani

Abstract

Biomimetic is the study of various principles of working mechanisms of naturally occurring phenomena and their further respective integrations in to such a modified advanced mechanized instruments/models of digital or artificial intelligence protocols. Hence, biomimetic has been proposed in last decades for betterment of human mankind for improving security systems by developing various convenient robotic vehicles and devices inspired by natural working phenomenon of plants, animals, birds and insects based on biochemical engineering and nanotechnology. Hence, biomimetic will be considered next generation technology to develop various robotic products in the fields of chemistry, medicine, material sciences, regenerative medicine and tissue engineering medicine, biomedical engineering to treat various diseases and congenital disorders. The characteristics of tissue engineered scaffolds are found to possess multifunctional cellular properties like biocompatibility, biodegradability and favorable mechanized properties when comes in close contact with the body fluids in vivo. This chapter will provide overall overview to the readers for the study based on reported data of developed biomimetic materials and tools exploited for various biomedical applications and tissue engineering applications which further helpful to meet the needs of the medicine and health care industries.

Keywords: Biomimicry, Biomimetic, Regenerative medicine, Tissue engineering

1. Introduction

The "biomimetics" originated from the Greek words "bios" (life) and "mimesis" (to imitate) and well known from ancient times to take ideas and inspirations from nature or surrounding environmental natural phenomena and various creatures like birds, animals, plants and insects for further transforming them in to most preferable and most promising practical and functional applications for the betterment of human mankind. Biomimetics consisted of new innovated urban design, innovative information technologies to facilitate understanding of the complex mechanisms of ecosystems and further followed with the mimicry of such systems in urban planning and management. Biomimetic is also called the most advanced the process of applying biological principles that underlie morphology, structures and functionality of biological entities to man-made design or models for the most efficient solution of existing challenges. For further proposed innovations, other relevant analysis of organisms and their respective ecosystems and biodiversity were also studied like their respective functions and associated processes based on exhibited scientific knowledge from biology and ecology. Most recently, biomimetic materials have been synthesized like Se-modified carbon nitride nanosheets, magnesium–strontium hydroxyapatite, dimethylglyoxime–urethane polyurethane, polydimethylsiloxane, Ag/Ag@AgCl/ZnO and PDTC(COOH)4/HA. Applications of biomimetic and biological materials are inevitable in various fields such as biomedical, oil–water separation, sensors, tissue engineering, genome technology

Research papers	Categorized under	Study focus	Highlights
1	Organism Level, Behavior Level, Ecosystem Level	Achieving more sustainability and **creating a regenerative built environment**	Framework is used to discuss various forms of biomimicry as a design methodology. (Realization process)
2	Ecosystem Level	Biomimicry's role in **addressing climate change** in the built environment	Certain ecosystem biomimicry principles for architecture have been deduced
3	Organism level	Principles have been **used in compression structures**	The potential use of tessellation and possibility of a modular building in the micro and macro scale
4	Organism level	**Breathing facades**, mimicking skin properties of flora and fauna	Enhancing cooling and ventilation systems and achieving thermal comfort in the buildings
5	Organism level	How termite mounds work and application of this concept in building design	A termite mound is a lung (Realization design, Object design)
6	Organism level	**Analyses mathematical patterns and structures** in biology and nature	A possibility of use of human bone in structure. Use of fiber reinforced concrete in building to achieve load bearing function.
7	Ecosystem Level	**Ecosystem based design theory**	A regenerative design approach (Realization process)
8	Ecosystem Level	A novel **optimization technique** inspired by natural ecosystems	Improving effectiveness of the user base
9	Organism level	Concept which **deals with the ability to collect water, sunlight and wind**	Outcome of Living building.
10	Ecosystem Level	**An ecological replacement scheme using traditional concepts** of natural landform, natural orientation and climatic resources.	Ecology principles in building design

Table 1.
Tabular documentation of key Highlights in biomimetic studies (reproduced by Desh SP, 2018) [14].

and ultrasound imaging [1]. Biomimetics has been proposed for developing various most novel nanotechnology technologies to find out many clinical and medical solutions to understand structural and functional properties of various biological components like proteins, amino acids and phospholipids to develop protein functionalized nanoparticles, peptide-functionalized gold nanoparticles, and carbohydrate-functionalized nanoparticles [1–3]. First, very well-known biomimetic based model named flying machine was invented by Leonardo da Vinci's (1452–1519) based on the most fundamental example of inspiration of birds to design "flying machine" and another named, "turtleship," a warship model was built to fight Japanese raiders during invasions [3–5]. The well-known, the Wright brothers (1867–1948) were also inspired to plot the note of the wings of eagles and made a powered airplane which succeeded in human flight for the first time in 1903 [6, 7]. A protein-driven nanocarrier device was composed of chemically modified nanocarrier consist of protein entities. Protein-based biomimetic nanocarriers were considered the effective biosafety carrier to be used in treatment of tumor having great success to carry out more effective targeted delivery of anticancer drugs and the gene therapy especially [8, 9]. Several modern clinical practices have been adopted to combat sudden increase in antimicrobial resisting bacteria with biomimetic strategies due to having inherent compatibility with physiologically relevant environments. The biomimetic based technologies have raised interest as an emerging field to have potential in treatment of tumor. Clinical Strategy has been proposed for combining nano-technology with biomimetic technology that has found to be gain increasing attention for developing more advanced bioinspired, environmentally benign, and promising diagnostic and therapeutic devices. And, developments of surgical needles had been done to make them safer by using biodegradable polymer and polylactic acid which significantly contribute for the advancement of biomimetics and biomedical engineering [10, 11]. Previously reported methods of fabrications of tissue engineering scaffolds were found to have many advantages over other conventional methods where biomaterials in micro/nano based surface modifications have chosen as for designing of biomimetic materials consisted of 3D printing and stem cells which have observed more effective for tissue engineering of bone and cartilage tissues [12, 13]. 3D printing has emerged as a critical biomimetics based fabrication process for bone engineering due having good control bulk geometry and internal structure of tissue scaffolds. Improved bioprinting methods and biocompatible ink materials for bone engineering have been observed potent optimal hybridized 3D scaffolds for bone defect repair including improved cellular function, cellular viability, mechanical integrity, biological activity, mechanical strength, easy fabrication and controllable degradation (**Table 1**). And, 3D printing might be helpful for next generation of bone grafts clinical practices to create on-demand patient-specific scaffolds [15, 16].

2. Clinical approaches

Antireflective coatings were developed by taking inspiration from phenomenon of moth's eyes called "Areflexia" which method involved refraction of the light significantly decreasing allows the moth to avoid predators and to see prey in the darkness. Hence, this robotic method is used for various military operations to develop the solar cell light-emitting diodes. Most interesting concept of biomimetic technology is to fabricate optical, electric and electronic properties of nanoparticles by controlling their size and shape by using simple preparatory protocols having less toxicity and trustworthy applications. Bio-inspired technology was found to most promising to develop biodegradable polymeric nanoparticles which can easily circulate through the blood for extended periods of time and also act as detoxification device [17, 18]. By mimicking red blood cells, both physical and chemical biomimicry are integrated to improve the biological function of nanomaterials in-vitro and in-vivo systems. The anisotropic shape and membrane coating are found to synergize for resisting the cellular uptake and reduce clearance from the blood. This approach is observed for enhancing the detoxification of nanoparticles and the anisotropic membrane-coated nanoparticles have enhanced biodistribution and therapeutic efficacy. These biomimetic biodegradable nanodevices and their derivatives have promise for applications ranging from detoxification agents, to drug delivery vehicles, and to biological sensors [19, 20]. High-strength carbon nanotubes (CNTs) were prepared by taking inspirations from "Mussels" called byssi composed of the crosslinking of collagen fibers and a protein known as Mefp-1. This innovative development of high strength was proposed in modern clinical approaches for stitching wounds and successful surgeries. In past decades, various biomimetic technologies have been opted to design the vaccine adjuvants to achieve better immune protection efficiency which could easily mimic the physical and chemical properties of natural pathogens when their interaction is taken place with antigen-presenting cells (APCs) (**Figure 1**).

By mimicking natural pathogens, these particles possess the ability to increase the cellular uptake of antigens, activate APCs, and promote the lysosomal escape of antigens [22]. Newly proposed carriers are synthesized with better biocompatibility, biodistribution and targeting characteristics by taking inspiration biological molecules, organisms, cells, drug vector performances. These prepared protocells are found to exhibited compartmentalized microarchitectures coded with the assembly of protein–polymer nanostructure conjugates used for encapsulation to articulate the proposed design of protein and its related enzymatic activities. The cell-derived biomimetic drug delivery systems are combined with most advantageous synthetic nanoparticles and a natural bio-membrane for providing enhanced efficacy in the treatment of atherosclerosis. The most considerable advantages of this combined system are to provide more options for the design of specific cell membrane coated nano-drugs for imparting good biocompatibility, long course time and inflammatory site targeting. These nanovehicles having protein-based biomimetic nanocarriers have many superior capabilities such as good biocompatibility, low toxicity, low side effects and improved chemotherapy effect. Newly innovated synthetic DNA-based nanoparticles have been synthesized by using biomimetic technology which are found to have full control over the trafficking of internal cargo cellular delivery of loaded compounds to targeting vectors which requires robust endosomal escape from the cellular degradation pathway to facilitate therapeutic mRNA or CRISPR/Cas9 action [23]. These days, the advanced drug delivery methods have been focused for achieving more successful targeted novel drug involving enhanced capacity of drug loading in drug carriers, cellular uptake of drug carriers and the sustained release of loaded drugs to the target cells. Six groups of therapeutic drug carriers are reported which including biomimetic hydrogels, biomimetic micelles, biomimetic liposomes, biomimetic dendrimers, biomimetic polymeric carriers and biomimetic nanostructures. Artificial chemically modified fabrication of biomimetic nanocomposite drug carriers could noticeably deliver the effective targeted concentration of loaded drugs in targeted drug delivery systems. Biomimetic hydrogels have emerged the most promising tissue engineering scaffold materials and their versatile chemistry can sum up various multiple physical and chemical features to integrate cells, scaffolds, and signaling molecules for tissue regeneration. Due to having highly hydrophilic nature hydrogels can design the nutrient-rich aqueous environments for cells and soluble regulatory molecules can be readily incorporated for achieving cell proliferation and differentiation. Novel in-situ

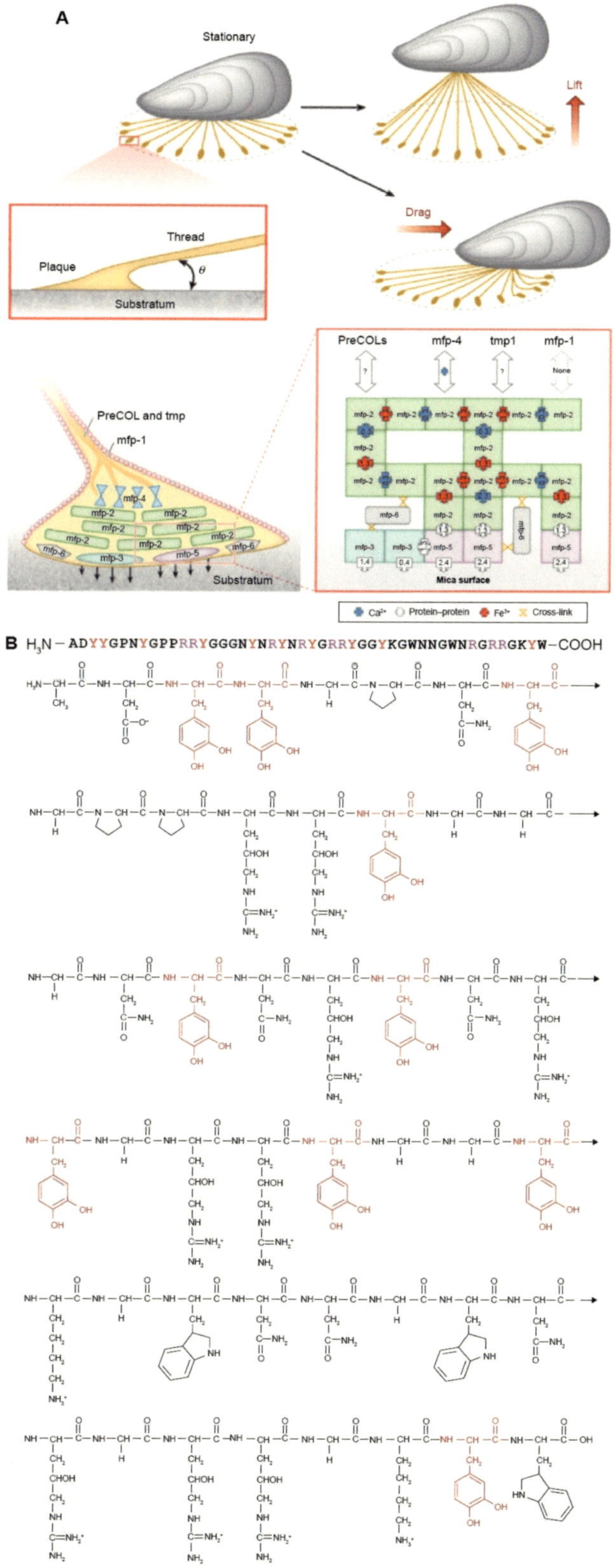

Figure 1.
(A) Structure of byssi in mussels. (B) Chemical structure of a Dopa-rich mussel foot protein (mfp) (reproduced by lee et al, 2011) [21].

chondroitin sulfate (CS) hydrogel was also synthesized by using phosphine-mediated Michael type addition reaction by adding precursor solutions of CS-acrylate and CS-tri(2-carboxyethyl)phosphine (TCEP) which was thermally stable and biocompatible [24]. The most important is their respective controlled dynamic parameters and spatial distribution of chemical signals in hydrogel scaffolds are critical for cell–cell communication, cell-scaffold interaction, and cell morphogenesis. Biomimetic hydrogels could be proposed for providing the supporting cells with spatiotemporally controlled chemical signals as tissue engineering scaffolds. These artificially designed hydrogels are found be helpful for clinically probing the temporally controlled growth factor-release abilities, spatially controlled conjugated bioactive molecules/motifs, and targeting delivery and reload properties for tissue engineering applications including exhibiting improved clinical characteristics like injectability, self-healing ability, stimulus-responsiveness and pro-remodeling features Multifunctional fibrous scaffolds have been also synthesized having high potential for bone regeneration which composed of poly d,l-lactide-co-trimethylene carbonate (PLMC) and worked as biomimicking attributes of poly d,l-lactideco-trimethylene carbonate nanofibers having improved efficacies and potency as scaffold materials for tissue repair and regeneration (**Figure 2**) [25]. Another new advanced technology based on cell membrane-covered or coated biomimetic nanovehicles for biomedical application has been seeking increasing attention involving membranes from red blood cells, platelets, leukocytes, tumor/cancer, and stem cells which are proposed as biomimetic coatings.

of nanoparticles for eluding the stimulated immune system to maintain their respective targeting capability. Biomimetic technology has also been incorporated in many robotics innovations to make robotic legs and feet for handicapped patients. Most recent, highly advanced biomimetic medical approach was reported for designing 5-degree-of-freedom robotic exoskeleton for upper limb therapy as most hi-tech rehabilitation robots by using CATIA software which inspired by the morphology of the bones and the muscle force transmission of the upper limbs [26, 27]. Scientists have been proposed various protocols for mimic the biological systems for achieving molecular scale control via self-assembly and directed assembly techniques via computerized fabrication and biochemical modification

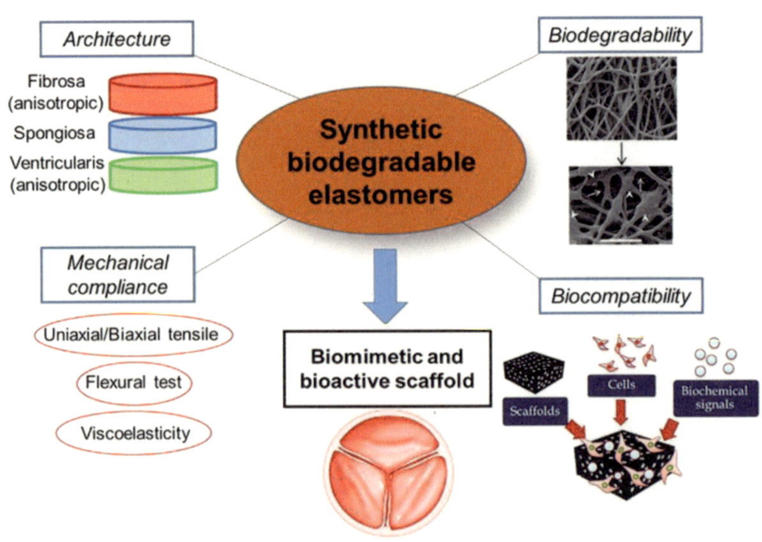

Figure 2.
Biodegradable and biomimetic elastomeric scaffolds (reproduced by Xue et al, 2017) [25].

with molecular precision to develop novel nanoscale devices to carry out diagnostic and therapeutic applications. Biomimetic technology has been adopted for designing and synthesis of a family of chiral and conformationally constrained involving preparation of a 160,000-member library of diverse tetramers via split-and-pool methods and from this library, a non-covalent ligand to the DNA-binding domain of p53 was invented for finding out its most feasible biomedical applications [27]. Nanozymes have been exploited in biomedicine and biomolecular detection which found to exhibit natural enzyme-mimicking catalytic activities to be used as specific nanocatalytic tumor therapy. The construction of an efficient biomimetic dual inorganic nanozyme-based nanoplatform is found to helpful for tumor microenvironment responsive nanocatalytic tumor therapy which triggers cascade catalytic reactions based on micro/submicron/nanosized Au and Fe^3O^4 NPs coloaded dendritic mesoporous silica nanoparticles. Various reported in-vitro and in-vivo clinical interpretations have found higher nanocatalytic-therapeutic efficient and safe with a desirable tumor-suppression rate (69.08%) based on their respective biocompatible composite nanocatalysts. Due to having observed high efficiency of in-vitro nanocatalytic therapy for killing cancer cells, the prolonged blood circulation, and potent tumor accumulation effect, the in vivo nanocatalytic therapeutic efficiency of DMSN-Au-Fe3O4 NPs was assessed against the 4 T1 breast tumor xenograft on nude mice. Hence, these proposed DMSN-Au-Fe^3O^4 composite nanosystems have also been evaluated for further clinical translation assessments. Therefore, proposed biomimetic dual inorganic nanocomposite-triggered cascade reaction strategy for TME-responsive and effective nanocatalytic tumor therapy is found to accepted paradigm of toxin-free-drug endogenous and noninvasive nanocatalytic biomedicine by adopting multienzyme mimicking catalytic activities for tumor-specific therapies [28, 29]. Recently, Inflammatory eye diseases including dry eye disease, uveitis, allergic conjunctivitis, scleritis, glaucoma, retinopathy and Age-Related Macular Degeneration (AMD) are treated by using biomimetic based ophthalmic drug delivery of immunomodulatory agents to improve vision-threatening ill effects in subjected patients as safe and easy controlled-release of loaded formulation with improved patient compliance and treatment efficacy. This proposed biomimetic ophthalmic drug delivery practice has pacifying symptoms in subjected patients with negligible dangerous side effects [30–33]. Three dimensional nanofibrous extracellular matrix mimicking structures are proposed to be found suitable to be used as tissue engineering scaffolds composed of biochemically modified natural and synthetic polymers to develop more advanced biomedical tools and devices which have potential of replication of the biomineralization and bone formation as well as tissue regeneration [34–36].

3. Conclusions

Hence, Biomimetic based technologies have been considered the most advanced alternative methods of chemically and bio-engineered drug delivery vehicles and devices. And, it needs lots of innovative inputs and related interpretations to make them more efficient and cost effective [36, 37]. Taken together, these findings indicate that biomimetics is becoming a dominant paradigm for robotics, materials science and other technological disciplines, with the potential for significant scientific, societal and economic impact over this decade and into the future. Hence, biomimicry is novel science stream which studies nature's models and further imitates or takes inspiration from these designs and their respective processes to solve human problems through engineering tools and artificial intelligence protocols (**Figure 3**) [37–39]. So, this innovative perspective fastened the scope of biomimetics (three

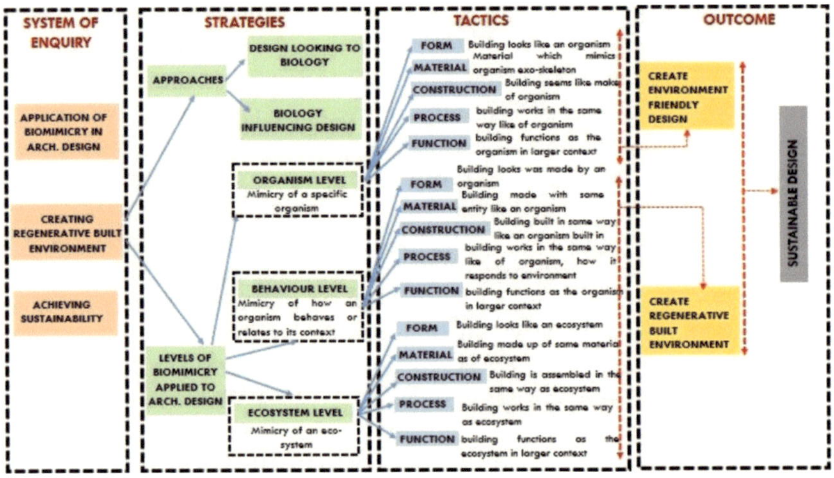

Figure 3.
Study mapping of biomimetics associated strategies (reproduced by dash SP, 2018) [14].

levels of biomimicry named, the organism level, behavior level and ecosystem level) to design its space to carry out most promising and novel solutions. So, environment inspiring biomimicry based robotic innovations could might have more potential to carry out more potent clinical and medical outcomes than any chemical and artificial alternative for developing more safe artificial intelligence technology-based products. And, biomimetic study can be evolved as more safe and efficient technology in future to develop human and environment friendly robotic products through integration with fields of applied chemistry, metabolomics, nano-technology, biomedical engineering [40–43]. Therefore, the developments of novel biomimicked biomaterials are observed for more responsive against stimulus could be considered the next choice to generate smart three dimensional biomimetic scaffolds that designed to perform more effective interaction with biological systems. So that, they can be used for a wide range of biomedical applications like delivery of loaded bioactive molecules and cell adhesion mediators to perform better cellular functioning to treat targeted diseases [44, 45].

Author details

Kirti Rani
Amity Institute of Biotechnology, Amity University Uttar Pradesh, Noida, UP, India

*Address all correspondence to: krsharma@amity.edu

References

[1] Speck O, Speck T. An overview of bioinspired and Biomimetic self-repairing materials. *Biomimetics*. 2019, 4(1): 26. https://doi.org/10.3390/biomimietics4010026.

[2] Kumar NS, Suvarna RP, Naidu KCB, Banerjee P, Ratnamala A, Manjunatha H. A review on biological and biomimetic materials and their applications. *Applied Physics A*. 2020; 126, 445.

[3] Perera AS and Coppens AO. Re-designing materials for biomedical applications: from biomimicry to nature-inspired chemical engineering. *Philosophical Transactions of the Royal Society A*. 2018. https://doi.org/10.1098/rsta.2018.0268.

[4] Shimomura M. The new trends in next generation biomimetics material technology: learning from biodiversity. *Sci Technol Trends Q Rev*. 2010; 37:53-75.

[5] Jakab PL. Visions of a Flying Machine: The Wright Brothers and the Process of Invention. *Smithsonian Institution Press*; 1990.

[6] Pemsel H. A History of War at Sea: An Atlas and Chronology of Conflict at Sea from Earliest Times to the Present. *Naval Institute Press*; 1977.

[7] Boxerbaum AS, Shaw KM, Chiel HJ, Quinn RD. Continuous wave peristaltic motion in a robot. *Int J Robot Res*. 2012;31(3):302-318.

[8] Freedman R. The Wright Brothers: How They Invented the Airplane. *New York: Holiday House*; 1991.

[9] Altendorfer R, Moore N, Komsuoglu H, *et al*. RHex: a biologically inspired hexapod runner. *Auton Robots*. 2001;11(3):207-213.

[10] Choi J, Hwang MP, Lee J-W, Lee KH. A glimpse into the interactions of cells in a microenvironment: the modulation of T cells by mesenchymal stem cells. *Int J Nanomedicine*. 2014;9(suppl 1):127.

[11] Li M, Huang X, Tang TD, Mann S. Synthetic cellularity based on nonlipid micro-compartments and protocell models. *Curr Opin Chem Biol*. 2014; 22:1-11.

[12] Chen, C., Bang, S., Cho, Y. et al. Research trends in biomimetic medical materials for tissue engineering: 3D bioprinting, surface modification, nano/micro-technology and clinical aspects in tissue engineering of cartilage and bone. *Biomater Res*. 2016; 20: 10 https://doi.org/10.1186/s40824-016-0057-3

[13] Lee KBL, Hui JHP, Song IC, Ardany L, Lee EH. Injectable mesenchymal stem cell therapy for large cartilage defects—A porcine model. *Stem Cells*. 2007; 25:2964-71.

[14] Dash, P. S. Application of Biomimicry in Building Design. *Int J Civil Engg Tech*. 2018, 9 (2), 644-660.

[15] Chia HN, Wu BM. Recent advances in 3D printing of biomaterials. J Biol Eng. 2015; 9:4.

[16] Zhang L, Yang G, Johnson BN, Jia X. Three-dimensional (3D) printed scaffold and material selection for bone repair. *Acta Biomater*. 2019, 15;84:16-33. doi: 10.1016/j.actbio.2018.11.039. Epub 2018 Nov 24. PMID: 30481607.

[17] Feng Y, Zhu S, Mei D, Li J, Zhang J, Yang S, Guan S. Application of 3D Printing Technology in Bone Tissue Engineering: A Review. *Curr Drug Deliv*. 2020 Nov 12. doi: 10.2174/156720181799 9201113100322. Epub ahead of print. PMID: 33191886.

[18] Wilke P, Helfricht N, Mark A, Papastavrou G, Faivre D, Börner HG.

A direct biocombinatorial strategy toward next generation, mussel glue inspired saltwater adhesives. *J Am Chem Soc*. 2014;136(36): 12667-12674.

[19] Hong S, You I, Song IT, Lee H. Material-independent surface modification inspired by mussel-adhesion. *Polym Sci Technol.* 2012; 23(4):396-406.

[20] HoonáKim B, YoungáKim J, OukáKim S. Directed self-assembly of block copolymers for universal nanopatterning. *Soft Matter.* 2013; 9(10):2780-2786.

[21] Li M, Huang X, Tang TD, Mann S. Synthetic cellularity based on nonlipid micro-compartments and protocell models. *Curr Opin Chem Biol*. 2014; 22:1-11.

[22] Lee BP, Messersmith PB, Israelachvili JN, Waite JH. Mussel-inspired adhesives and coatings. *Annual review of materials research*. 2011, 41:99.

[23] Bao M, Lou X, Zhou Q, Dong W, Yuan H, Zhang Y. Electrospun biomimetic fibrous scaffold from shape memory polymer of PDLLAco-TMC for bone tissue engineering. *ACS Appl Mater Interfaces*. 2014; 6(4): 2611-2621.

[24] Gong Y-K, Winnik FM. Strategies in biomimetic surface engineering of nanoparticles for biomedical applications. *Nanoscale*. 2012;4(2): 360-368.

[25] Jo S, Kim S, Noh I. Synthesis of in situ chondroitin sulfate hydrogel through phosphine-mediated Michael type addition reaction. *Macromol Res*. 2012; 20: 968-76.

[26] Xue Y, Sant V, Phillippi J, Sant S. Biodegradable and biomimetic elastomeric scaffolds for tissue-engineered heart valves. *Acta Biomaterialia*, 2017; 48, 2-19. https://doi.org/10.1016/j.actbio.2016.10.032.

[27] Kim Y-C, Park J-H, Prausnitz MR. Microneedles for drug and vaccine delivery. *Adv Drug Deliv Rev*. 2012;64(14):1547-1568.

[28] Paul Dominick E. Baniqued, Jade R. Dungao, Michael V. Manguerra, Renann G. Baldovino, Alexander C. Abad, and Nilo T. Bugtai.. Biomimetics in the design of a robotic exoskeleton for upper limb therapy. *AIP Conference Proceedings*, 2018; 1933, 040006. https://doi.org/10.1063/1.5023976.

[29] Aquino C, Sarkar M, Michael J. Chalmers, Mendes K, Kodadek T, Glenn C. Micalizio. A biomimetic polyketide-inspired approach to small-molecule ligand discovery. *Nature Chemistry*, 2012; 4: 99-104. doi:10.1038/nchem.1200

[30] Gao S, Lin H, Zhang H, Yao H, Chen Y, Shi J, Nanocatalytic Tumor Therapy by Biomimetic Dual Inorganic Nanozyme-Catalyzed Cascade Reaction. *Advance Science,* 28 November 2018. https://doi.org/10.1002/advs.201801733

[31] Ratay ML, Balmert SC, Acharya AP, Greene AC, Meyyappan T, Little SR. Controlled release of an HDAC inhibitor for reduction of inflammation in dry eye disease. *Acta Biomater*. 2018; 71:261-270. doi: 10.1016/j.actbio.2018.03.002.

[32] Hwang, Jangsun & Jeong, Yoon & Park, Jeong & Lee, Kwan & Cho, Dong-Woo & Choi, Jonghoon. (2015). Biomimetics: Forecasting the future of science, engineering, and medicine. International journal of nanomedicine. 10. 5701-5713. 10.2147/IJN.S83642.

[33] Ratay ML, Bellotti E, Gottardi R, Little SR. Modern Therapeutic Approaches for Noninfectious Ocular Diseases Involving Inflammation. *Adv Healthc Mater*. 2017; 6(23). doi: 10.1002/adhm.201700733.

[34] Reddy R, Reddy N. Biomimetic approaches for tissue engineering.

J Biomater Sci Polym Ed. 2018; 29(14):1667-1685. doi: 10.1080/09205063.2018.1500084. Epub 2018 Sep 26. PMID: 29998794.

[35] Khan F, Tanaka M. Designing Smart Biomaterials for Tissue Engineering. *Int J Mol Sci*. 2017 Dec 21;19(1):17. doi: 10.3390/ijms19010017. PMID: 29267207; PMCID: PMC5795968.

[36] Ravichandran R, Sundarrajan S, Venugopal JR, Mukherjee S, Ramakrishna S. Advances in polymeric systems for tissue engineering and biomedical applications. *Macromol Biosci*. 2012 Mar;12(3):286-311. doi: 10.1002/mabi.201100325. Epub 2012 Jan 25. PMID: 22278779.

[37] Zare, M.; Falahat, M. Characteristics of reptiles as a model for bionic architecture. *Adv. Civ. Environ. Eng*. 2013, 1, 124-135.

[38] Lopez, M.; Rubio, R.; Martin, S.; Croxford, B. How plants inspire facades. From plants to architecture: Biomimetic principles for the development of adaptive architectural envelopes. *Renew. Sustain. Energy Rev*. 2017, 67, 692-703.

[39] Buck, N.T. The art of imitating life: The potential contribution of biomimicry in shaping the future of our cities. *Environ. Plan. B Urban Anal. City Sci*. 2017, 44, 120-140.

[40] Al-Obaidi, K.M.; Ismail, M.A.; Hussein, H.; Rahman, A.M.A. Biomimetic building skins: An adaptive approach. *Renew. Sustain. Energy Rev.* 2017, 79, 1472-1491.

[41] Elmeligy, D.A. Biomimetics for ecologically sustainable design in architecture: A proposed methodo logical study. Eco-Archit. VI Harmon. *Archit. Nat*. 2016, 161, 45-57.

[42] Fujii, S.; Sawada, S.; Nakayama, S.; Kappl, M.; Ueno, K.; Shitajima, K.; Butt, H.-J.; Nakamura, Y. Pressure-sensitive adhesive powder. *Mater. Horiz*. 2016, 3, 47-52.

[43] Khelil, S.; Zemmouri, N. Biomimetic: A new strategy for a passive sustainable ventilation system design in hot and arid regions. *Int. J. Environ. Sci. Technol*. 2019, 16, 2821-2830.

[44] Gao, R.; Liu, K.; Li, A.; Fang, Z.; Yang, Z.; Cong, B. Biomimetic duct tee for reducing the local resistance of a ventilation and air-conditioning system. *Build. Environ*. 2018, 129, 130-141.

[45] Liu Z, Li H, Cao G. Quick Estimation Model for the Concentration of Indoor Airborne Culturable Bacteria: An Application of Machine Learning. *Int J Environ Res Public Health*. 2017;14(8): 857. doi: 10.3390/ijerph14080857. PMID: 28758941; PMCID: PMC5580561.